Nonparametric Estimation of Probability Densities and Regression Curves

Mathematics and Its Applications (*Soviet Series*)

E. A. Nadaraya
Tbilisi State University, Tbilisi, U.S.S.R.

Nonparametric Estimation of Probability Densities and Regression Curves

Translated by
Samuel Kotz

Kluwer Academic Publishers
Dordrecht / Boston / London

Library of Congress Cataloging in Publication Data

Nadaraya, E. A., 1935-
Nonparametric estimation of probability densities and regression curves.

(Mathematics and its applications. Soviet series)
Translated from the Russian.
Includes index.
1. Distribution (Probability theory) 2. Estimation theory. 3. Nonparametric statistics. 4. Regression analysis. I. Title. II. Series: Mathematics and its applications (D. Reidel Publishing Company). Soviet series.
QA273.6.N33 1989 519.2 88-12647
ISBN 90-277-2757-0

Published by Kluwer Academic Publishers
P.O. Box 17, 3300 AA Dordrecht, The Netherlands.

Kluwer Academic Publishers incorporates the publishing programmes of D. Reidel, Martinus Nijhoff, Dr W. Junk and MTP Press.

Sold and distributed in the U.S.A. and Canada by Kluwer Academic Publishers,
101 Philip Drive, Norwell, MA 02061, U.S.A.

In all other countries, sold and distributed by Kluwer Academic Publishers Group,
P.O. Box 322, 3300 AH Dordrecht, The Netherlands

This is a translation of the original work: Непараметрическое Оценивание Плотности Вероятностей и Кривой Регрессии
Published by Tbilisi University Press

Printed in The Netherlands

SERIES EDITOR'S PREFACE

'Et moi, ..., si j'avait su comment en revenir, je n'y serais point allé.'

Jules Verne

The series is divergent; therefore we may be able to do something with it.

O. Heaviside

One service mathematics has rendered the human race. It has put common sense back where it belongs, on the topmost shelf next to the dusty canister labelled 'discarded nonsense'.

Eric T. Bell

Mathematics is a tool for thought. A highly necessary tool in a world where both feedback and nonlinearities abound. Similarly, all kinds of parts of mathematics serve as tools for other parts and for other sciences.

Applying a simple rewriting rule to the quote on the right above one finds such statements as: 'One service topology has rendered mathematical physics ...'; 'One service logic has rendered computer science ...'; 'One service category theory has rendered mathematics ...'. All arguably true. And all statements obtainable this way form part of the raison d'être of this series.

This series, *Mathematics and Its Applications*, started in 1977. Now that over one hundred volumes have appeared it seems opportune to reexamine its scope. At the time I wrote

> "Growing specialization and diversification have brought a host of monographs and textbooks on increasingly specialized topics. However, the 'tree' of knowledge of mathematics and related fields does not grow only by putting forth new branches. It also happens, quite often in fact, that branches which were thought to be completely disparate are suddenly seen to be related. Further, the kind and level of sophistication of mathematics applied in various sciences has changed drastically in recent years: measure theory is used (non-trivially) in regional and theoretical economics; algebraic geometry interacts with physics; the Minkowsky lemma, coding theory and the structure of water meet one another in packing and covering theory; quantum fields, crystal defects and mathematical programming profit from homotopy theory; Lie algebras are relevant to filtering; and prediction and electrical engineering can use Stein spaces. And in addition to this there are such new emerging subdisciplines as 'experimental mathematics', 'CFD', 'completely integrable systems', 'chaos, synergetics and large-scale order', which are almost impossible to fit into the existing classification schemes. They draw upon widely different sections of mathematics."

By and large, all this still applies today. It is still true that at first sight mathematics seems rather fragmented and that to find, see, and exploit the deeper underlying interrelations more effort is needed and so are books that can help mathematicians and scientists do so. Accordingly MIA will continue to try to make such books available.

If anything, the description I gave in 1977 is now an understatement. To the examples of interaction areas one should add string theory where Riemann surfaces, algebraic geometry, modular functions, knots, quantum field theory, Kac-Moody algebras, monstrous moonshine (and more) all come together. And to the examples of things which can be usefully applied let me add the topic 'finite geometry'; a combination of words which sounds like it might not even exist, let alone be applicable. And yet it is being applied: to statistics via designs, to radar/sonar detection arrays (via finite projective planes), and to bus connections of VLSI chips (via difference sets). There seems to be no part of (so-called pure) mathematics that is not in immediate danger of being applied. And, accordingly, the applied mathematician needs to be aware of much more. Besides analysis and numerics, the traditional workhorses, he may need all kinds of combinatorics, algebra, probability, and so on.

In addition, the applied scientist needs to cope increasingly with the nonlinear world and the extra mathematical sophistication that this requires. For that is where the rewards are. Linear models are honest and a bit sad and depressing: proportional efforts and results. It is in the non-linear world that infinitesimal inputs may result in macroscopic outputs (or vice versa). To appreciate what I am hinting at: if electronics were linear we would have no fun with transistors and computers; we would have no TV; in fact you would not be reading these lines.

There is also no safety in ignoring such outlandish things as nonstandard analysis, superspace and anticommuting integration, p-adic and ultrametric space. All three have applications in both electrical engineering and physics. Once, complex numbers were equally outlandish, but they frequently proved the shortest path between 'real' results. Similarly, the first two topics named have already provided a number of 'wormhole' paths. There is no telling where all this is leading - fortunately.

Thus the original scope of the series, which for various (sound) reasons now comprisese five sub-series: white (Japan), yellow (China), red (USSR), blue (Eastern Europe), and green (everything else), still applies. It has been enlarged a bit to include books treating of the tools from one subdiscipline which are used in others. Thus the series still aims at books dealing with:

- a central concept which plays an important role in several different mathematical and/or scientific specialization areas;
- new applications of the results and ideas from one area of scientific endeavour into another;
- influences which the results, problems and concepts of one field of enquiry have, and have had, on the development of another.

The present volume is concerned with nonparametric questions in statistics. Traditional (parametric) statistics deals with methods and problems in estimation on the basic assumption that the underlying statistical probability distribution is one of a family depending on just a few parameters. This is an assumption which is often violated and is, in fact, in many circumstances not a very reasonable assumption. That is where nonparametric and robust methods come in. Here, roughly, robust statistics assumes that the true distribution is in some (small) neighbourhood of a parametric family and nonparametric statistics makes almost no assumptions at all beyond, say, as in this book, smoothness of the underlying distribution.

This book is about nonparametric statistics; more specially it is about direct estimation of the probability density; still more specifically it is about the powerful techniques and ideas developed rather recently both in the USSR (by the author among others) and in the West which make effective use of the assumption that the underlying distribution is smooth.

Nonparametric statistics is, of course, very widely applicable. It also generates, as this book shows, very interesting mathematics; a combination which makes this work most attractive for this series.

Perusing the present volume is not guaranteed to turn you into an instant expert, but it will help, though perhaps only in the sense of the last quote on the right below.

The shortest path between two truths in the real domain passes through the complex domain.

J. Hadamard

La physique ne nous donne pas seulement l'occasion de résoudre des problèmes ... elle nous fait pressentir la solution.

H. Poincaré

Never lend books, for no one ever returns them; the only books I have in my library are books that other folk have lent me.

Anatole France

The function of an expert is not to be more right than other people, but to be wrong for more sophisticated reasons.

David Butler

Bussum, October 1988 Michiel Hazewinkel

TABLE OF CONTENTS

Series Editor's Preface v

Preface to the English Edition ix

Introduction 1

CHAPTER 1. ASYMPTOTIC PROPERTIES OF CERTAIN MEASURES OF DEVIATION FOR KERNEL-TYPE NONPARAMETRIC ESTIMATORS OF PROBABILITY DENSITIES 18

1. Integrated Mean Square Error of Nonparametric Kernel-Type Probability Density Estimators 18
2. The Mean Square Error of Nonparametric Kernel-Type Density Estimators 30

CHAPTER 2. STRONGLY CONSISTENT IN FUNCTIONAL METRICS ESTIMATORS OF PROBABILITY DENSITY 42

1. Strong Consistency of Kernel-Type Density Estimators in the Norm of the Space C 42
2. Convergence in the L_2 Norm of Kernel-Type Density Estimators 47
3. Convergence in Variation of Kernel-Type Density Estimators and its Application to a Nonparametric Estimator of Bayesian Risk in a Classification Problem 54

CHAPTER 3. LIMITING DISTRIBUTIONS OF DEVIATIONS OF KERNEL-TYPE DENSITY ESTIMATORS 62

1. Limiting Distribution of Maximal Deviation of Kernel-Type Estimators 62
2. Limiting Distribution of Quadratic Deviation of Two Nonparametric Kernel-Type Density Estimators 87
3. The Asymptotic Power of the $U_{n_1 n_2}$-Test in the Case of 'Singular' Close Alternatives 99
4. Testing for Symmetry of a Distribution 103
5. Independence of Tests Based on Kernel-Type Density Estimators 109

CHAPTER 4. NONPARAMETRIC ESTIMATION OF THE REGRESSION CURVE AND COMPONENTS OF A CONVOLUTION 115

1. Some Asymptotic Properties of Nonparametric Estimators of Regression Curves 115
2. Strong Consistency of Regression Curve Estimators in the Norm of the Space C(a, b) 122
3. Limiting Distribution of the Maximal Deviation of Estimators of Regression Curves 125
4. Limiting Distribution of Quadratic Deviation of Estimators of Regression Curves 138
5. Nonparametric Estimators of Components of a Convolution (S.N. Bernstein's Problem) 152

CHAPTER 5. PROJECTION TYPE NONPARAMETRIC ESTIMATION OF PROBABILITY DENSITY 161

1. Consistency of Projection-Type Probability Density Estimator in the Norms of Spaces C and L_2 161
2. Limiting Distribution of the Squared Norm of a Projection-Type Density Estimator 164

Addendum LIMITING DISTRIBUTION OF QUADRATIC DEVIATION FOR A WIDE CLASS OF PROBABILITY DENSITY ESTIMATORS 177

1. Limiting Distribution of U_n 178
2. Kernel Density Estimators / Rosenblatt-Parzen Estimators 186
3. Projection Estimators of Probability Density / Chentsov Estimators 194
4. Histogram 199
5. Deviation of Kernel Estimators in the Sence of the Hellinger Distance 200

References 204

Author Index 210

Subject Index 212

PREFACE TO THE ENGLISH EDITION

This monograph is a modest attempt to acquaint the reader with an area of mathematics which can be described as "curve estimation". This includes, for example, estimation of density functions, regression functions, etc. This is a very important area of mathematical statistics which is being intensively developed at present and in which many papers are being published every year. It is not the intention that this book seeks to present a comprehensive overview of the literature.

The general purpose and plan of the book remains the same as that of the 1983 Russian edition. However, this English edition includes additional material in which the limit distribution of the quadratic deviation for a broad class of density function estimators and also the limit distribution of the measure of closeness in the sense of the Hellinger distance for kernel density function estimators is studied.

I'd like to express my deep gratitude to the staff of Kluwer Academic Publishers for their effective cooperation in realizing this English translation.

I am grateful to Professor S. Kotz for the excellent translation of my book.

Finally, it is a pleasure to thank the assistant to the Publisher Ms. O.A. Pols for her efficient help.

March 28, 1988
Tbilisi

E.A. Nadaraya

INTRODUCTION

The theory of statistical estimation is one of the basic branches of mathematical statistics. This theory is subdivided into parametric and nonparametric estimation. A nonparametric procedure is usually defined as a procedure which is valid independently of the distribution of the sampled observations. The problem of estimating the functional characteristics of a distribution law of observations belongs to problems of nonparametric estimation. In particular, in recent years there has been a growing interest in problems of estimating probability density and regression curves. This monograph is devoted to a study of these problems.

We note that problems of accuracy of estimation of a probability density were originally posed and studied in the Soviet Union. The results obtained in this area are due to V. I. Glivenko and N. V. Smirnov who used histograms for estimators. V. I. Glivenko [1] established almost sure (a.s.) uniform convergence of a histogram to a continuous theoretical density. Important results in the area have been obtained by N. V. Smirnov [2] who derived the limiting distribution for the maximum of the absolute value of a normed deviation of a histogram from theoretical smooth density. Smirnov's results were subsequently generalized by S. Kh. Tumanyan [3].

M. Rosenblatt [4], E. Parzen [5] and N. N. Chentsov [6] made further contributions to the theory of nonparametric estimation of probability densities. In their works, these authors introduced new classes of estimators which generalize histograms. The idea of constructing new nonparametric density estimators is as follows. Let $X_1, X_2, \ldots, X_n$ be a sequence of independent identically distributed random variables with the distribution law $\vec{P}(\cdot)$. The empirical distribution $\vec{P}_n(\cdot)$ constructed from the sample $X_1, \ldots, X_n$ is a discrete distribution with atoms of weight $1/n$ located at each one of the observed points. Associated with each observation X_i is a delta-measure $\delta_{X_i}(\cdot)$ concentrated at point X_i and with a sequence of independent observations $X_1, X_2, \ldots, X_n$ - the arithmetic mean of measures

$$\delta_{X_i}(\cdot), \quad \text{i.e.} \quad \hat{P}_n(\cdot) = 1/n \sum_{i=1}^{n} \delta_{X_i}(\cdot).$$

If the unknown distribution possesses a smooth density $f(x)$, it is natural to 'spread out' each observed measure $\delta_{X_i}(\cdot)$ replacing it by a measure with a certain density $K_n(x, X_i)$ and to use the Radon-Nikodym derivatives of such a 'spread out' distribution as an estimator of the density $f(x)$. We thus arrive at the following class of estimators:

$$f_n(x) = 1/n \sum_{i=1}^{n} K_n(x, X_i). \tag{0.1}$$

The choice of the function $K_n(x, y)$ is quite important and, to a large extent, determines the properties of $f_n(x)$. One class of estimators - the so-called 'kernel' estimator - was proposed by M. Rosenblatt [4] in 1956 and E. Parzen [5] in 1962. Another class of estimators was also introduced in 1962 by N. N. Chentsov [6] who called them 'projection' estimators in orthogonal series.

At present, a number of other approaches to non-parametric density estimation [7, 8] is available in world statistical literature. We shall not discuss them but proceed to a brief description of Rosenblatt's, Parzen's and Chentsov's results which have a direct relation to this monograph.

An interesting result - due to M. Rosenblatt [4] - deals with the problem of unbiasedness of density estimators. Let $S(y; x_1, \ldots, x_n)$ be a non-negative Borel measurable function with respect to $(y, x_1, x_2, \ldots, x_n)$. Then $S(y; X_1, \ldots, X_n)$ cannot be an unbiased estimator of $f(y)$ for all continuous $f(y)$ and for all y.

Rosenblatt's [4] class of estimators is generated by the kernel

$$K_n(x, y) = a_n K(a_n(x - y)),$$

i.e. estimators in this class are of the form

$$f_n(x) = \frac{a_n}{n} \sum_{i=1}^{n} K(a_n(x - X_i)), \tag{0.2}$$

where $\lim_{n\to\infty} a_n = \infty$ and $K(x)$ is a function which satisfies certain regularity conditions. Rosenblatt has devoted special attention to the case: $K(x) = \frac{1}{2}$ (= 0) for $|x| \le 1$ ($|x| > 1$). For this case, he investigated the asymptotics of the mean square error $E(f_n(x) - f(x))^2$. G. M. Maniya [8] generalized Rosenblatt's results to the multi-dimensional case.

E. Parzen [5] continued the study of estimators of type (0.2). In particular, he established asymptotic unbiasedness and asymptotic normality of such estimators. He obtained an asymptotic expression for the bias and the mean square deviation. Furthermore, he introduced the estimator Θ_n of the mode Θ of a distribution defined by $f_n(\Theta_n) = \max_x f_n(x)$ and proved the consistency and asymptotic normality of Θ_n in the case when Θ is unique.

N. N. Chentsov [6] proposed another class of estimators which he called projection estimators. The basic idea of his approach is to approximate an unknown density $f(x) \in L_2(r)$ ($r(x)$ is an arbitrary non-negative function) by a segment of its Fourier series in terms of an appropriate system of orthonormal functions $\phi_1(x), \phi_2(x), \ldots$ The class of Chentsov's [6] estimators is generated by the kernel $K_N(x, y) = \sum_1^n \phi_j(x)\phi_j(y)r(y)$, where $N = O(n)$, i.e. the estimators are of the form:

$$f_n(x) = \sum_{j=1}^{N} \hat{a}_j\phi_j(x), \tag{0.3}$$

where

$$\hat{a}_j = \frac{1}{n}\sum_{k=1}^{n} \phi_j(X_k)r(X_k).$$

We present an interesting result due to Chentsov [6]. Under certain restrictions

$$\rho(\| f_n - f \|) \asymp [n^{-1}\Gamma(n)]^{1/2},$$

where $\|\cdot\|$ is the norm of the space $L_2(r)$. The function Γ is defined as $\Gamma(y/g(y)) \equiv y$ and $g(y) \downarrow 0$ as $y \uparrow \infty$ and $\rho(\|\cdot\|)$ is either $(E\|f_n - f\|^2)^{1/2}$ or the quantile $Q(\|f_n - f\|)$ of level $1 - \delta$, $0 < \delta < 1$.

Thus, in the earlier papers [4]-[6], discussed above, a number of specific problems associated with estimation of probability density were studied.

Naturally it becomes necessary to develop a general theory which would permit us to approach the investigation of a wide variety of problems associated with estimation of various functional characteristics of distributions of observations from a unified position. Such a theory will be developed in this monograph. Here, general methods for an asymptotic analysis of nonparametric estimators of a wide class of functional characteristics of a distribution are developed. These methods allow us to investigate problems of constructing asymptotically optimal procedures of estimation, confidence regions and tests for testing statistical hypotheses.

In this monograph, sequences of asymptotically optimal (in various senses) estimators of a number of functional characteristics of a distribution of observations are obtained, and convergence - in certain probabilistic senses - of the constructed estimators in functional metrics C, L_2 and L_1 is investigated, and correctness in the sense of convergence in variation of the statistical problem of estimating an unknown absolutely continuous distribution is proved; a method is developed for investigating the limiting distribution for the maximum of the absolute value of normalized deviation of estimators of the type (0.2) from an unknown density. The latter allows us to construct confidence regions for the unknown density. Moreover, new tests of the integral type, based on density estimators investigated herein for testing goodness of fit, homogeneity, symmetry and independence are developed. Here, for the first time, a sequence of 'close' alternatives, which we call 'singular' is introduced. Under these alternatives, the proposed tests turn out to be more powerful than the classical tests of the Kolmogorov-Smirnov type. Nonparametric estimators of regression curves are introduced for the first time; problems of asymptotic behavior of deviations (in the integral and uniform metrics) of these estimators from the theoretical regression curve are investigated, and nonparametric confidence regions for the regression curve are constructed. Tests for testing the appropriate statistical hypotheses are developed and a solution of the well-known S. N. Bernstein's problem of estimating a density with errors is obtained.

The results and methods presented in this monograph have generated interest from other researchers in the USSR as well as abroad (cf. for example [9], [10], [11], [12], etc.).

In this monograph the usual dual enumeration of theorems and formulas is used separately in each section. When referring to a theorem or a formula in another chapter, the chapter number is stated first. For example (4.3.5) indicates a reference to Formula 5 in Section 3 of Chapter 4.

The book consists of five chapters.

The first chapter entitled: 'Asymptotic properties of certain measures of deviation for kernel-type nonparametric estimators of probability densities' consists of two sections. These sections are devoted to the study of asymptotics of two measures of closeness of estimators from the class (0.2): the mean value of the integral of squared error (m.i.s.e.) and the mean quadratic deviation.

The first result of Section 1 in Chapter 1 consists of the derivation of an asymptotic expansion for the m.i.s.e.

$$U(a_n) = E\int [f_n(x) - f(x)]^2\, dx.$$

(Below if the domain of integration is not indicated it will mean that integration is over the whole space). If $f(x) \in W_s$ and $K(x) \in H_s$ (see Section 1 of Chapter 1 for the definitions of these classes of functions), then

$$U(a_n) = \frac{a_n}{n}\int K^2(x)\, dx + a_n^{-2s}\frac{\alpha^2}{(s!)^2}\int [f^{(s)}(x)]^2\, dx + {} + O\left(\frac{a_n}{n} + a_n^{-2s}\right), \qquad (0.4)$$

where

$$\alpha = \int x^s K(x)\, dx \neq 0 \quad \text{and} \quad a_n/n \to 0 \quad \text{as} \quad n \to \infty.$$

The optimal sequence $\{a_n^0\}$ for which $U(a_n)$ possesses the best order of magnitude is:

$$a_n^0 = \Theta_1(f)n^{1/(1+2s)}$$

and for the optimal a_n^0

$$U(a_n^0) = \Theta_2(f)n^{-2s/(1+2s)} + o\left(n^{-2s/(1+2s)}\right),$$

where Θ_1 and Θ_2 are functionals on $f(x)$ (Θ_1 and Θ_2 are written explicitly in (1.1.11)).

Theorem 1.1.2 presents a generalization of (0.4) to the multidimensional case.

The optimal sequence $\{a_n^0\}$ depends on the unknown density. In this connection it is both of interest and necessity for practical purposes to obtain - based on a sample - estimators of the terms of an optimal sequence for which the m.i.s.e. approaches zero 'in the same manner' as for the optimal sequence. The second basic result in Section 1 of Chapter 1 deals with this problem. A sequence of estimators $\hat{a}_n^0 = \Theta_n(X_1, X_2, \ldots, X_n)$ of the terms a_n^0 is constructed (cf. (1.1.13)) for which it is proved (Theorem 1.1.3) that

$$U(\hat{a}_n^0) = U(a_n^0) + o\left(n^{-2s/(1+2s)}\right). \tag{0.5}$$

At the end of the section, the expectation $E\|f_n - f\|_p^p$, where $p \geq 2$ is even is considered as a measure of error of the estimator (0.2) ($\|\cdot\|_p$ is the norm of the space $L_p(-\infty, \infty)$) and it is shown (Theorem 1.1.4) that

$$\limsup_n n^{ps/(1+2s)} E\|f_n - f\|_p^p \leq C(f, K, s, p),$$

where $C(f, K, s, p)$ is a constant.

In the second section estimators $f_n(x)$ of the type (0.2) are studied for a multivariate density $f(\vec{x})$, $\vec{x} = (x_1, x_2, \ldots, x_p)$ (cf. (1.2.1)). Here an asymptotic expansion of the quantity $\eta^2(f_n) = E(f_n(\vec{x}) - f(\vec{x}))^2$ is obtained (Theorem 1.2.1); for a special choice of the parameter a_n it is shown that

$$\eta^2(f_n) \sim \Theta_3(f)n^{-2s/(p+2s)}. \tag{0.6}$$

Farrel [13] has shown that the order in (0.6) cannot be improved. Theorem (1.2.3) solves the problem analogous to (0.5) which represents a generalization of Theorem 4.1 of paper [14]. Next, certain asymptotic properties of estimators of partial derivatives of a probability density are studied (Theorem 1.2.2).

Chapter 2 entitled 'Strongly consistent in functional metrics estimators of probability density' consists of three sections. In this chapter, convergence in certain probabilistic senses of nonparametric estimators of the type (0.2) in functional metrics C, L_2 and L_1 is studied.

In Section 1 the problem of deviations of estimators (0.2) from f(x) in the uniform metric is considered. The main aim of this section is to solve the following problem (due to N. V. Smirnov): under what conditions on f(x), K(x) and a_n does $f_n(x)$ converge almost surely (a.s.) to f(x) uniformly on the whole real line? Theorem 2.1.1 is devoted to a solution of this problem. The theorem presents sufficient conditions under which $\sup_x |f_n(x) - f(x)| \to 0$ a.s. as $n \to \infty$. This theorem also implies that the sample mode Θ_n defined above converges to the mode Θ of the distribution a.s. (Theorem 2.1.2). This result was generalized to the multidimensional case by Van Ryzin [15] and Vaduva [16]. It is necessary to note that the conditions of our Theorem 2.1.1 in a certain sense cannot be improved. Schuster [17] has shown that if K(x) and a_n satisfy the conditions of Theorem 2.1.1 then the uniform continuity of f(x) on the whole axis is a necessary condition for the uniform convergence of $f_n(x)$ to f(x) a.s. In Subsection 2 of Section 1 of Chapter 2 a brief proof of Schuster's theorem is presented. At the end of the section a generalization of Theorem 2.1.1 to the multidimensional case is given.

In the subsequent two sections multidimensional estimators $f_n(\vec{x})$, $\vec{x} \in R_p$, (R_p is a p-dimensional Euclidean space) of the type (0.2) are studied. In Section 2 of Chapter 2 the norm in the space $L_2(R_p)$ of the deviation of $f_n(\vec{x})$ from the true $f(\vec{x})$: $||f_n - f||_2$ is used as a measure of the error of the estimator $f_n(\vec{x})$. Sufficient conditions are presented under which $||f_n - f||_2 \to 0$ a.s. as $n \to \infty$ (Theorem 2.2.1). In problems of nonparametric inference (cf. Subsection 2 of Section 2 in Chapter 2) the functional $\alpha(f)$ plays an important role. This functional is equal to the integral of $f^2(x)$ and is usually unknown. It is thus natural to use $\alpha(f_n)$ as an estimator of $\alpha(f)$. It follows from Theorem 2.2.1 that $\alpha(f_n) \to \alpha(f)$ a.s. as $n \to \infty$ (Theorem 2.2.2).

Let $P_n\{\cdot\}$ and $P\{\cdot\}$ be probability measures on the σ-algebra B_p of Borel sets in R_p corresponding to $f_n(\vec{x})$ and $f(\vec{x})$. As a measure of error of the estimator $P_n\{\cdot\}$ we use the variation of deviation of the measure $P_n\{\cdot\}$ from the true probability measure $P\{\cdot\}$:

$$||P_n - P|| = \sup_{A \in B_p} |P_n\{A\} - P\{A\}| .$$

In Section 3 of Chapter 2 the following basic problem is solved: is there a convergence of the sequence of measures $P_n\{\cdot\}$, n = 1, 2, ..., to the measure $P\{\cdot\}$ in variation in a certain probabilistic sense without any assumptions on the density $f(x)$? Theorems 2.3.1, 2.3.2 and 2.3.5. provide an answer to this problem. Specifically Theorems 2.3.1 and 2.3.2 provide a class of kernels $K(x)$ and sequences a_n for which $||P_n - P|| \to 0$ a.s. for an arbitrary density $f(x)$. In Theorem 2.3.5 an analogous class of $K(x)$ and of a_n which assure the convergence $||P_n - P|| \to 0$ in probability is presented. Theorems 2.3.1, 2.3.2 and 2.3.5 provide a refinement of the assertion stated by N. N. Chentsov in his lecture [11].

The convergence of measures $P_n\{\cdot\}$ to $P\{\cdot\}$ in variation has applications in classification problems (cf. Subsection 2 of Section 3 in Chapter 2). Namely, in Section 2 the strong consistency of an estimator of a Bayesian risk of a classification is studied (Theorem 2.3.3). The empirical region of classification constructed in this connection 'approaches' asymptotically in a certain metric to the unknown optimal set.

The third chapter entitled 'Limiting distributions of deviations of kernel-type density estimators' consists of five sections.

In Section 1 of Chapter 3 we consider problems associated with the limiting distribution of the maximal deviation of an estimator $f_n(x)$ of type (0.2) from the unknown density $f(x)$ on a finite interval $[a, b]$ (N. V. Smirnov's problem). The theorems proved allow us to carry out statistical estimation of the degree of approximation and to construct for $f(x)$ a confidence region with a given confidence coefficient. It is well known that in N. V. Smirnov's groundbreaking paper [2] the limiting distribution of the maximal deviation for the case when $f_n(x)$ is a 'histogram' has been derived.

The problem of determining the limiting distribution of the maximal deviation of the estimator (0.2) is subdivided into two cases: a 'regular' and a 'singular' one.

A 'regular' is the case when $f(x) \geq \mu > 0$ on a finite interval $[a, b]$. 'Singular' is the case when $f(x)$ may vanish at certain isolated points of the interval or to approach zero at infinity in the limit.

In Subsections 2-6 of Section 1 of Chapter 3 the singular case is investigated. In Subsection 2 the notation and basic assumptions are presented. These assumptions are on the density of observations $f(x)$, the kernel $K(x)$ and a_n. In Subsection 3 the basic Lemma 3.1.3 is proved which deals with the asymptotic behavior of probabilities of large deviations for $\xi_n(x)$ where $\xi_n(x)$ is a normalized value of the estimator $f_n(x)$. Let E_i, $i = \overline{1, s}$ be finite and non-overlapping intervals in $R = (-\infty, \infty)$ and $E_{s+1} = R \setminus \bigcup_{k=1}^{s} E_k$ and $s = s(n)$ as $n \to \infty$. Denote by Δ_{s_n} the sum of lengths of these intervals and $\mu_{s_n} = \min f(x)$, $x \in R \setminus E_{s_n+1}$.

Lemma 3.1.3 asserts that

$$P\{|\xi_n(t_j)| \geq \lambda_{s_n}, \quad j = \overline{1, r}\} = \left(\frac{2}{\sqrt{2\pi}} \int_{\lambda_{s_n}}^{\infty} \exp\left(-\frac{x^2}{2}\right) dx\right)^r \times$$

$$\times \left[1 + O\left(\frac{\Delta s_n}{s_n \mu_{s_n}} \lambda^2_{s_n} \right) + O\left(\frac{\sqrt{s_n}\, \lambda^m_{s_n}}{\sqrt{n \mu_{s_n}}\, \Delta_{s_n}} \right) \right],$$

where t_j are the centers of the intervals E_j and λ_{s_n} is a non-negative function increasing in s_n and $m = \max(3, r)$.

In Subsection 4 auxiliary assertions are presented. In Subsection 5, limit theorems on maximal deviation are proved. The theorems obtained applied to the construction of asymptotic confidence regions for the unknown density of observations $f(x)$. As an example, we present one result from Subsection 5 (Theorem 3.1.4):

$$\lim_{n\to\infty} P\left\{ \sqrt{na_n^{-1}} \max_{x \in R \setminus E_{s_{n+1}}} |T_n(x) - f(x)| f^{-1/2}(x) \leq \right.$$
$$\left. \leq K_0 \left(\ell_{s_n} + \frac{\lambda}{\ell_{s_n}} \right) \right\} = S(\lambda) = e^{-2e^{-\lambda}},$$

where

$$T_n(x) = f_n(t_i) \text{ for } x \in E_i, \quad K_0^2 = \int K^2(x)\, dx$$

and ℓ_{s_n} is the root of the equation $1/s_n = 1 - \Phi(\ell)$ ($\Phi(\ell)$ is the standard normal distribution function). In Subsection 6 examples are presented which clarify the meaning of Theorems 3.1.1 and 3.1.4.

In Subsection 7 the problem of the maximal deviation of $f_n(x)$ from $f(x)$ in the regular case is considered. Here we present one of the results obtained (Theorem 3.1.7):

$$\lim_{n\to\infty} P\left\{ \sqrt{na_n^{-1}} \max_{x \in [a,b]} |\tilde{f}_n(x) - f(x)| f^{-1/2}(x) < K_0 \left(\ell_{s_n} + \frac{\lambda}{\ell_{s_n}} \right) \right\} =$$
$$= S(\lambda), \tag{0.7}$$

where

$$\tilde{f}_n(x) = f(t_k) + (x - t_k) a_n^{-1} (f_n(t_{k+1}) - f_n(t_k)) \quad \text{for}$$
$$x \in [t_k, t_{k+1}].$$

At the end of Section 1 of Chapter 3 the problem of minimizing a confidence region for $f(x)$, constructed with the aid of Theorems 3.1.4 and 3.1.7, is considered. We note that Theorems 3.1.4 and 3.1.7 contain the

Smirnov classical results [2] which we derive under less restrictive conditions.

Let two independent samples of sizes n_1 and n_2 be given from populations with densities $f_1(x)$ and $f_2(x)$ respectively. On the basis of these samples we construct two non-parametric estimators $f_{n_1}(x)$ and $f_{n_2}(x)$ respectively. On the basis of these samples we construct two non-parametric estimators $f_{n_1}(x)$ and $f_{n_2}(x)$ of type (0.2) with parameters $a_1 = a_1(n_1)$ and $a_2 = a_2(n_2)$.

Section 2 of Chapter 3 is devoted to a determination of the limiting distribution of the quadratic deviation between $f_{n_1}(x)$ and $f_{n_2}(x)$:

$$U_{n_1n_2} = \frac{N_1N_2}{N_1 + N_2} \int (f_{n_1}(x) - f_{n_2}(x))^2 r(x)\, dx, \quad N_i = n_i a_i^{-1},$$

$$i = 1, 2.$$

Let $f_1(x) = f_2(x) = f_0(x)$ (hypothesis H_0), $f_0(x)$ be the given density and let the sample sizes increase in such a manner that $n_2/n_1 = \tau$ and $a_1 = a_2 = a_n$, $n = n_2/(1 + \tau)$.

Set

$$A(f_0) = \int f_0(x) r(x)\, dx \int K^2(u)\, du,$$

$$\sigma^2(f_0) = 2 \int f_0^2(x) r^2(x)\, dx \int K_1^2(u)\, du, \quad K_1 = K*K.$$

Under the conditions of Theorem 3.2.2 $a_n^{1/2}(U_{n_1n_2} - A(f_0))$ is distributed asymptotically normally $(0, \sigma(f_0))$. Theorem 3.2.2. allows us to construct an asymptotic test at level α for testing the hypothesis H_0. Here H_0 is rejected if

$$U_{n_1n_2} \geq d_n(\alpha) = A(f_0) + a_n^{-1/2}\lambda_\alpha \sigma(f_0).$$

It follows from this theorem (cf. corollary of Theorem 3.2.2) that the random variable

$$a_n^{1/2}(U_{n_1n_2} - A(f_{n_1n_2}))\sigma^{-1}(f_{n_1n_2}),$$

where

$$f_{n_1n_2}(x) = (n_1 + n_2)^{-1}(n_1 f_{n_1}(x) + n_2 f_{n_2}(x)),$$

is asymptotically normally distributed with parameters (0, 1). Therefore one can construct a test for testing the hypothesis $H_0 : f_1(x) = f_2(x)$. The critical region is determined by the inequality

$$U_{n_1 n_2} \geq \tilde{d}_n(\alpha) = A(f_{n_1 n_2}) + a_n^{-1/2} \lambda_\alpha \sigma(f_{n_1 n_2}).$$

Next, local behavior of the power of these tests for the case of a sequence of alternatives 'close' to the null hypothesis is considered. The sequence is of the form

$$H_n : f_i(x) = f_0(x) + n^{-1/2+\delta/4} \phi_i(x), \quad i = 1, 2$$

$$1/4 < \delta < 1/2. \tag{0.8}$$

It is shown (Theorem 3.2.3) that

$$\lim P_{H_n}(U_{n_1 n_2} \geq D_n(\alpha)) = 1 - \Phi\left(\lambda_\alpha - \frac{1}{\sigma} \int (\phi_1(x) - \phi_2(x))^2 r(x)\, dx\right),$$

where

$$D_n(\alpha) = d_n(\alpha) \quad \text{or} \quad D_n(\alpha) = \tilde{d}_n(\alpha).$$

From here it follows that for positive weight functions r(x) the test for testing H_0 versus H_n is asymptotically strictly unbiased. At the end of the section an example of comparing the power of the $U_{n_1 n_2}$-test with the corresponding powers of $\chi^2_{n_1 n_2}$ and $V^2_{n_1 n_2}$-tests [18] is presented. It is shown that the $U_{n_1 n_2}$-test is more powerful than both $\chi^2_{n_1 n_2}$-tests relative to the alternative H_n.

The aim of Section 3 of Chapter 3 is to compare the tests based on estimators of distribution functions and densities. As it is known for the tests constructed in terms of empirical distribution functions local 'close' alternatives differ from the null hypothesis by a quantity of order $n^{-1/2}$. Therefore it is clear that for alternatives similar to (0.8) the tests based on $U_{n_1 n_2}$ are less powerful than, for example, tests of the Kolmogorov-Smirnov-Cramér-von Mises type.

In this section new types of 'converging' alternatives called 'singular' are constructed for which tests based on $U_{n_1 n_2}$ will be more powerful than tests of the type indicated above since the latter tests cannot distinguish between hypotheses which converge faster than $n^{-1/2}$.

A sequence of 'singular' alternatives is of the form

$$H_n : f_1(x) = f_0(x), \quad f_2(x) = f_0(x) + \alpha_n \phi_n(x - x_0), \tag{0.9}$$

where $\phi_n(x)$ satisfy certain regularity conditions, $\alpha_n \downarrow 0$ in an appropriate manner and x_0 is a fixed point of continuity of $r(x)$ such that $r(x_0) \neq 0$.

The limiting power function of the $U_{n_1 n_2}$-test under alternatives (0.9) is obtained in Theorem 3.3.1 and its corollary. Thus

$$\lim P_{H_n}(U_{n_1 n_2} \geq D_n(\alpha)) = 1 - \Phi(\lambda_\alpha - C_0 r(x_0)\sigma^{-1}),$$

where

$$D_n(\alpha) = d_n(\alpha) \quad \text{or} \quad D_n(\alpha) = \tilde{d}_n(\alpha), \quad 0 < C_0 = \lim_{n\to\infty} n a_n^{-1/2} \alpha_n^2 \beta_n,$$

$$\beta_n = \int \phi_n^2(x)\, dx = O(1), \quad \alpha_n \cdot \beta_n = O(n^{-1/2}) \quad \text{and}$$

$$\int |\phi_n(x)|\, dx = O(\beta_n).$$

Section 4 of Chapter 3 is devoted to the investigation of the limiting distribution of the statistic

$$T_n = n a_n^{-1}\ 1/2 \int [f_n(x) - f_n(-x)]^2 r(x)\, dx.$$

Under the hypothesis H_0 according to which $f(x) = f(-x)$ for all $x \in (-\infty, \infty)$ it is asserted (Theorem 3.4.1) that the random variable $a_n^{1/2}(T_n - A(f_n))\sigma^{-1}(f_n)$ is distributed asymptotically normally (0.1). Next, local behavior of the test with the critical region $T_n \geq \tilde{d}_n(\alpha) = A(f_n) + \lambda_\alpha \sigma(f_n) a_n^{-1/2}$ (Theorems 3.4.2 and 3.4.3) in the case of two ((0.7) and (0.8)) types of converging alternatives. In particular, in Theorem 3.4.3 the particular case (0.9) is discussed, i.e. the case when $\phi_n(x) = W(x/\beta_n)$, $\beta_n \downarrow 0$ in an appropriate manner, $W(x) \neq W(-x)$ and $\alpha_n \cdot \beta_n = O(n^{-1/2})$. It is proved in the theorem that

$$\lim P_{H_n}(T_n \geq \tilde{d}_n(\alpha)) = 1 - \Phi\left(\lambda_\alpha - \frac{r(0)}{\sigma} \int (W(x) - W(-x))^2\, dx\right).$$

Hence tests based on deviations between empirical distribution functions as, for example, a test of the ω_n^2-type [19] and tests of the Kolmogorov-Smirnov type [20] cannot distinguish the 'singular' hypothesis from the null hypothesis when the power of the T_n-test tends to a nondegenerate limit, i.e. the power tends to a number which is larger than the level of the tests.

Let a sample of size n from a population with density $f(\vec{x}) = f_1(x_1)\cdot f_2(x_2)$ (hypothesis H_0) be given, $\vec{x} = (x_1, x_2)$. In Section 5 of Chapter 3 the limiting distribution of the functional

$$S_n = na_n^{-2} \int (f_n(\vec{x}) - f_{n_1}(x_1)f_{n_2}(x_2))^2\, r(\vec{x})\, d\vec{x},$$

is studied; here $f_n(\vec{x})$, $f_{n_1}(x)$, $f_{n_2}(x)$ are estimators of the type (0.2) of the densities $f(\vec{x})$, $f_1(x)$ and $f_2(x)$ respectively.

Denote

$$A_n = \int f_{n_1}(x_1)f_{n_2}(x_2)r(\vec{x})\, d\vec{x} \int K^2(u)\, du, \qquad K(\vec{u}) = K_1(u_1)K_2(u_2),$$

$$\sigma_n^2 = \int f^2_{n_1}(x_1)f^2_{n_2}(x_2)r^2(\vec{x})\, d\vec{x} \int K^2_{10}(x_1)\, dx_1 \int K^2_{20}(x_2)\, dx_2,$$

$$K_{i0} = K_i * K_i, \qquad i = 1,2.$$

It is proved (Theorems 3.5.2 and 3.5.3) that the random variable $a_n\sigma_n^{-1}(S_n - A_n)$ is distributed asymptotically normally (0.1). Next a test for testing the hypothesis $H_0 : f(\vec{x}) = f_1(x_1)\cdot f_2(x_2)$ is constructed; the critical region for testing this hypothesis is of the form $S_n \geq A_n + a_n^{-1}\sigma_n\lambda_\alpha$.

Chapter 4 entitled 'Nonparametric estimation of regression curves and components of a convolution' consists of five sections.

In Sections 1-4 the asymptotic properties of a non-parametric estimator $r_n(x)$ of the regression curve $r(x)$ which was introduced and studied by the author of this monograph in [21] are investigated. As it was already mentioned, many statisticians have been attracted to an investigation of the properties of density estimators (0.2). Investigations of the properties of $r_n(x)$ are at least equally as popular. In this direction many interesting results were obtained both in the USSR and abroad (cf., e.g., [10], [16], [22], [23], [24] and others).

Let (X_i, Y_i), $i = \overline{1, n}$ be a sample from a two-dimensional random variable (X, Y). Denote the regression function of Y onto X by $r(x)$, $r(x) = E(Y|X = x)$.

As an approximation to the regression curve $r(x)$ the following class of statistics is considered

$$r_n(x) = \begin{cases} \dfrac{\sum_{i=1}^{n} Y_i K(a_n(x - X_i))}{\sum_{i=1}^{n} K(a_n(x - X_i))}, & \text{if } \sum_{i=1}^{n} K(a_n(x - X_i)) \neq 0 \\ 0, & \text{if } \sum_{i=1}^{n} K(a_n(x - X_i)) = 0. \end{cases} \tag{0.10}$$

In Subsection 1 of Section 1 of Chapter 4 conditions are studied under which the estimators (0.10) are consistent (Theorem 4.1.1), asymptotically normal (Theorem 4.1.3). Also the behavior of the bias is investigated (Theorem 4.1.2). Theorem 4.1.4 presents an asymptotic expansion for $r_n(x)$. Next, based on the results of Theorem 4.1.3 a confidence interval for $r(x)$ is constructed (cf. (4.1.20)).

In the second subsection of Section 1 of Chapter 4 the quantity

$$U_n = E \int (r_n(x) - r(x))^2 g_n^2(x) h(x)\, dx$$

is introduced. This quantity is used for the study of an error of the estimator $r_n(x)$. Here $g_n(x)$ is an estimator of the type (0.2) of the density $g(x)$ of the random variable X and $h(x)$ is a weight function. The best order of decrease of U_n towards 0 is obtained (Theorem 4.1.5).

In Section 2 of Chapter 4 the problem of deviation of estimators (0.10) from $r(x)$ in the uniform metric is studied. Sufficient conditions (Theorem 4.2.1) for convergence in $C(a, b)$ a.s. of estimators (0.10) to the theoretical $r(x)$ are presented.

In Section 3 of Chapter 4 limiting theorems are obtained for the maximal deviation of the nonparametric estimator of a regression curve $r_n(x)$ from the theoretical $r(x)$.

The problem of obtaining the limiting distribution of the maximal deviation of the estimator $r_n(x)$ from $r(x)$ on a finite interval has been solved and then utilizing the same ideas Theorem 4.3.3 was extended to the multidimensional case [24].

In Subsection 2 of Section 3 assumptions and auxiliary lemmas are presented. Here also the main Theorem 4.3.3 is proved according to which

$$\lim_{n\to\infty} P\left\{\sqrt{\frac{n}{a_n}} \max_i \left|\frac{r_n(t_i) - r(t_i)}{\sigma(t_i)}\right| < \left(\ell_{s_n} + \frac{\lambda}{\ell_{s_n}}\right) K_0\right\} = S(\lambda),$$

where

$$\sigma^2(x) = \beta(x)/g(x), \quad \beta(x) = D(Y/X = x), \quad \text{and} \quad \ell_{s_n}, \ t_i \quad \text{and} \quad S(\lambda)$$

are defined as in (0.7)

In Subsection 3 of Section 3 based on Theorem 4.3.3, Theorems 4.3.4 and 4.3.6. are proved. These theorems assert that

$$\lim_{n\to\infty} P\left\{\max_{a\le x\le b} \left|\frac{m_n(x) - r(x)}{\sigma(x)}\right| < \left(\ell_{s_n} + \frac{\lambda}{\ell_{s_n}}\right) K_0\sqrt{\frac{a_n}{n}}\right\} = S(\lambda), \quad (0.11)$$

$$\lim_{n\to\infty} P\left\{\sqrt{\frac{n}{a_n}} \max_{a\le x\le b} \left|\frac{\tilde{m}_n(x) - r(x)}{\sigma(x)}\right| < \left(\ell_{s_n} + \frac{\lambda}{\ell_{s_n}}\right) K_0\right\} = S(\lambda), \quad (0.12)$$

respectively, where

$$m_n(x) = r_n(t_j) \quad \text{for} \quad x \in E_j \quad \text{and} \quad \tilde{m}_n(x) = r_n(t_j) +$$

$$+ a_n^{-1}(x - t_j)(r_n(t_{j+1}) - r_n(t_j)) \quad \text{for} \quad x \in E_j,$$

$$j = \overline{1, s_n}$$

In Theorem 4.3.8 with the aid of the results obtained, the two basic problems are solved: (1) the construction of an asymptotic confidence region for $r(x)$; (2) the construction of tests for testing the null hypothesis $H_0 : r(x) = r_0(x)$ ($r_0(x)$ is a given function). In the simplest case, namely for a linear model corresponding to a normal distribution the solution of problems 1 and 2 is well known.

A solution of these problems in the general case presented herein was initially given by the author in [24].

Section 4 of Chapter 4 is devoted to the investigation of the limiting behavior of the distribution of the quadratric deviation

$$W_n = na_n^{-1} \int (r_n(x) - r(x))^2 g_n^2(x) h(x) \, dx.$$

In Subsection 1 of Section 4 Theorem 4.4.1 is proved. This theorem states that the random variable

$$\sqrt{a_n}(W_n - \int \alpha(x)h(x) \, dx \int K^2(u) \, du)$$

is asymptotically normal with mean 0 and variance

$$\sigma^2 = 2 \int \alpha^2(x)h^2(x) \, dx \int K_1^2(u) \, du,$$

where

$$\alpha(x) = D(Y|X = x)g(x) \quad \text{and} \quad K_1 = K*K.$$

In Subsection 2 of Section 4 some statistical problems associated with $r(x)$ are considered. Namely, utilizing the result of Theorem 4.4.1 it is possible to construct tests for testing the hypothesis $H_0 : r(x) = r_0(x)$.

(1) If $\alpha(x)$ is known, then the critical region for testing the hypothesis H_0 is of the form $W_n \geq d_n(\alpha)$ where

$$d_n(\alpha) = \int \alpha(x)h(x) \, dx \int K^2(u) \, du + a_n^{-1/2}\lambda_\alpha\sigma.$$

(2) If, however, $\alpha(x)$ is unknown then the critical region for testing the hypothesis H_0 will be of the form $W_n \geq \tilde{d}_n(\alpha)$ (Theorem 4.4.2) where $\tilde{d}_n(\alpha)$ is obtained from $d_n(\alpha)$ by replacing $\alpha(x)$ with the corresponding consistent estimator $\alpha_n(x)$. Next, a sequence of 'close' alternatives H_n of the form (0.8) is considered and it is shown (Theorem 4.4.3) that

$$\lim_{n\to\infty} P\{W_n \geq D_n(\alpha)\} = 1 - \Phi(\lambda_\alpha - \sigma^{-1} \int \bar{r}^2(x) g^2(x) h(x)\, dx),$$

where

$$D_n(\alpha) = d_n(\alpha) \quad \text{or} \quad D_n(\alpha) = \tilde{d}_n(\alpha),$$

and $\bar{r}(x)$ is determined by the hypothesis H_n.

In Subsection 3 of Section 4 the problem of testing hypotheses concerning equality of regression functions is considered.

Let two independent samples $(X_j^{(i)}, Y_j^{(i)})$, $j = \overline{1, n_i}$, $i = 1, 2$, from populations $(X^{(i)}, Y^{(i)})$, $i = 1, 2$ be given. Furthermore, let $r_i(x)$, $i = 1, 2$ be the corresponding regression curves. Assume that the marginal densities of X's are the same and that

$$D(Y^{(1)} | X^{(1)} = x) = D(Y^{(2)} | X^{(2)} = x).$$

For testing the hypothesis $H_0 : r_1(x) = r_2(x)$ the statistic

$$L_{n_1 n_2} = \frac{N_1 N_2}{N_1 + N_2} \int (\phi_{n_1}(x) - \phi_{n_2}(x))^2 h(x)\, dx, \qquad N_i = n_i a_{n_i}^{-1}$$

is considered where

$$\phi_{n_i}(x) = \frac{a_{n_i}}{n_i} \sum_{j=1}^{n_i} Y_j^{(i)} K(a_{n_i}(x - X_j^{(i)})).$$

Under H_0 the statistic $a_n^{1/2}(L_{n_1 n_2} - \Delta)$ is distributed asymptotically normally $(0, \sigma)$, where Δ and σ are defined in (4.4.43). This proposition allows us to construct a test for testing the hypothesis $H_0 : r_1(x) = r_2(x)$. By means of this test the hypothesis H_0 is rejected if $L_{n_1 n_2} \geq \Delta_{n_1 n_2} + a_n^{-1/2} \sigma_{n_1 n_2} \lambda_\alpha$ where $\Delta_{n_1 n_2}$ and $\sigma_{n_1 n_2}$ are defined in (4.4.44).

In Section 5 of Chapter 4 we are dealing with the situation where values of an unknown random variable X are observed. This variable represents a sum of two independent variables Y and Z; Z is normally distributed with parameters $(0, \sigma)$ where σ is assumed to be known. It is the variable Y that we are concerned with since X may be viewed as an error in the measurement of Y. In this section two basic problems are solved: (1) estimation of the density function $g(x)$ of the variable Y by means of observed values $X_1, X_2, \ldots, X_n$ of the variable X (the so-called S. N. Bernstein problem [25]). (2) estimation, using the same data, of the regression function of Y onto X, $r(x) = E(Y | X = x)$.

Subsections 1-2 of Section 5 are auxiliary. Here properties of estimators of derivatives of a density and some corollaries to Rooney's theorem [26] dealing with Gauss' transformation are presented.

In Subsection 3-4 two types of estimators of density $g(x)$ are constructed. One is for the narrow class of kernels $K(x)$ and the other for a wider class. In Theorems 4.5.4 and 4.5.6 problems of unbiasedness and of consistency of these estimators is studied.

In Subsection 5 of Chapter 4 the second problem is solved. An estimator $r_n(x)$ for $r(x)$ is constructed and its asymptotic behavior is studied. In particular in Theorem 4.5.8 the asymptotic normality of $r_n(x)$ is proved. By means of this theorem a confidence interval for $r(x)$ is constructed. At the conclusion of Section 5 Theorem 4.3.9 is presented which generalizes Theorem 4.5.8.

Chapter 5 entitled 'Projection-type nonparametric estimation of probability density' consists of two sections.

In Section 1 of this chapter a brief survey of results obtained in this direction is presented as well as certain asymptotic properties of projection estimators (0.3) are studied. These properties include convergence to zero of the mean value of the square of a norm of the error and uniform convergence a.s.

In Subsection 2 of Section 2 of this chapter the problem of the limiting distribution for the square of the norm of the error of projection estimator (0.3) is considered.

Set

$$T_{n,N} = \| f_n - f \|_2^2, \quad \alpha_j(x) = \phi_j(x) r(x),$$

$$\Delta_N = \frac{1}{N} \sum_{j=1}^{N} \int \alpha_j^2(x) f(x)\, dx$$

$$\sigma_N^2 = \frac{2}{N} \sum_{i=1}^{N} \sum_{j=1}^{N} \left(\int \alpha_i(x) \alpha_j(x) f(x)\, dx \right)^2.$$

It is asserted in Theorem 5.2.1 that the distribution of the random variable $N^{1/2}(T_{n,N} - \Delta)$ converges to a normal distribution with mean 0 and variance σ^2 where $\Delta = \lim_{N\to\infty} \Delta_N$ and $\sigma^2 = \lim_{N\to\infty} \sigma_N^2$. In particular if $\{\phi_j\}$ is a system of trogonometric functions on $[-\pi, \pi]$ then Δ and σ possess explicit expressions (cf (5.2.13) and (5.2.14)).

In Subsection 2 of Section 2 some examples of the application of Theorem 5.2.1 are presented.

In Subsection 3 of this section the asymptotic behavior of the goodness of fit test based on $T_{n,N}$ is studied. Let the hypothesis $H_0: f(x) = f_0(x)$ concerning a sequence of 'close' alternatives H_n of the type (0.8) be tested. It turns out (Theorem 5.2.2) that

$$\lim P_{H_n}\left(T_{n,N} \geq \Delta_0 + \frac{\lambda_\alpha}{\sqrt{N}} \sigma_0 \right) = 1 - \Phi\left(\lambda_\alpha - \sigma_0^{-1} \int \phi^2(x) r(x)\, dx \right),$$

where Δ_0 and σ_0 are determined under H_0 and $\phi(x)$ - under H_n.

At the end of Section 2 from Theorem 5.2.2 the limiting power of the χ_n^2-test under the alternatives of the type (0.8) is derived as a particular case for the case of equiprobable groups (the number of groups increases with n).

It is with a feeling of great gratitude that we recall the name of a most prominent scholar in the field of probability theory and mathematical statistics, a corresponding member of the USSR Academy of Sciences, Professor N. V. Smirnov under whose guidance the author studied the theory of nonparametric estimation.

Conversations and numerous discussions with N. N. Chentsov have had substantial influence on this monograph.

I am deeply indebted to him for his unfailing interest in my investigations.

In conclusion I would like to thank I. A. Melamed who read the manuscript of this book and made a number of valuable comments.

Chapter 1

ASYMPTOTIC PROPERTIES OF CERTAIN MEASURES OF DEVIATION FOR KERNEL-TYPE NONPARAMETRIC ESTIMATORS OF PROBABILITY DENSITIES

1. Integrated Mean Square Error of Nonparametric Kernel-Type Probability Density Estimators

1. Let $X_1, X_2, \ldots, X_n$ be independent identically distributed one-dimensional random variables possessing unknown probability density function f(x). We assume that the required density f(x) belongs to the space $L_2(-\infty, \infty)$ of functions which are square integrable with respect to the Lebesgue measure and consider the methods of empirical approximations to this density - when the errors are measured in the $L(-\infty, \infty)$ metric - of the form

$$f_n(x, a_n) = n^{-1} a_n \sum_{i=i}^{n} K(a_n(x - X_i)), \tag{1.1}$$

Here K(x) is a function belonging to $L_2(-\infty, \infty)$ which satisfies the regularity conditions H_s stipulated below and a_n is a sequence of positive numbers converging to infinity but fulfilling the condition $a_n = o(n)$ as $n \to \infty$.

Various numerical indices determining the measure of deviation between $f_n(x, a_n)$ and f(x) were introduced and studied by many authors (cf. e.g. [4], [27]). In particular, the mean integrated squared error (m.i.s.e.) is such a measure:

$$U(a_n) = E \int [f_n(x, a_n) - f(x)]^2 \, dx. \tag{1.2}$$

The first result in Section 1 is the derivation of an asymptotic expansion of the m.i.s.e. and is a substantiation and generalization of the corresponding result by M. Rosenblatt [4] - which was obtained for a simple kernel K(x) - to a wide class of kernels. Nonparametric density estimation and consequently the smallness of the m.i.s.e. depend on some increasing sequence of numbers chosen by a statistician. An optimal sequence, i.e. one for which the m.i.s.e. possesses the best order of smallness depends on the probability density (which is unknown to the investigator). In this connection it is of interest and necessity for practical purposes to obtain - based on a sample - estimates of the members of the optimal sequence for which the m.i.s.e. approaches zero 'in the same manner' as in the case of an optimal sequence. This constitutes the second basic result of Section 1.

2. In this subsection we shall study the asymptotic behavior of the m.i.s.e. $U(a_n)$ as $n \to \infty$, thus generalizing Rosenblatt's result [4].

Denote by W_s the set of functions $\phi(x)$ possessing derivatives up to the s-th ($s \geq 2$) order inclusively and let the s-th derivative $\phi^{(s)}(x)$ be a continuous bounded function belonging to $L_2(-\infty, \infty)$.

A function $K(x)$ will be called a function belonging to the class H_s ($s \geq 2$ is an even number) if it satisfies the following regularity conditions:

$$K(x) = K(-x), \quad \int K(x)\,dx = 1, \quad \sup_{-\infty<x<\infty} |K(x)| \leq A < \infty, \tag{1.3}$$

$$\int x^i K(x)\,dx = 0, \quad i = \overline{1, s-1}, \quad \int x^s K(x)\,dx \neq 0,$$

$$\int x^s |K(x)|\,dx < \infty. \tag{1.4}$$

Functions presented in the table of paper [5] belong to H_2, Bartlett's function [28] defined by $K(x) = 9/8.\ (1 - \frac{5}{3}x^2)$ for $|x| \leq 1$ and $K(x) = 0$ for $|x| > 1$ belongs to H_4, while the function

$$K(x) = 15/8\left(1 - \frac{2}{3}x^2 + \frac{1}{15}x^4\right)\frac{1}{\sqrt{2\pi}}\exp\left\{-\frac{x^2}{2}\right\}$$

is a function belonging to H_6 and so on.

THEOREM 1.1. If $f(x) \in W_s$ and $K(x) \in H_s$ then as $n \to \infty$

$$U(a_n) = \frac{a_n}{n}\int K^2(x)\,dx + a_n^{-2s}\frac{\alpha^2}{(s!)^2}\int [f^{(s)}(x)]^2\,dx + + 0\left(\frac{a_n}{n} + a_n^{-2s}\right)$$

where

$$\alpha = \int x^s K(x)\,dx \neq 0. \tag{1.5}$$

<u>Proof</u>. Utilizing Fubini's theorem and (1.1) we have

$$U(a_n) = \int Df_n(x, a_n)\,dx + \int [Ef_n(x, a_n) - f(x)]^2\,dx,$$

where

$$\int Df_n(x, a_n)\,dx = \frac{a_n}{n}\int Ea_nK^2(a_n(x - X_1))\,dx - - n^{-1}\int [Ea_nK(a_n(x - X))]^2\,dx.$$

From here in view of Hölder's inequality and Fubini's theorem we have

$$\int Df_n(x, a_n)\,dx = \frac{a_n}{n}\int K^2(x)\,dx + 0\left(\frac{a_n}{n}\right). \tag{1.6}$$

Next, since $f(x) \in W_s$ and $K(x) \in H_s$ we can write

$$\int [Ef_n(x, a_n) - f(x)]^2\, dx = \frac{a_n^{-2s}}{((s-1)!)^2} \times$$
$$\times \int \left(\int\int_0^1 (1-t)^{s-1} f^{(s)}\left(x + t\frac{u}{a_n}\right) u^s K(u)\, du\, dt \right)^2 dx.$$

It thus remains to show that

$$A_n = \int \left(\int\int_0^1 (1-t)^{s-1} f^{(s)}\left(x + t\frac{u}{a_n}\right) u^s K(u)\, du\, dt \right)^2 dx \to$$
$$\to \frac{1}{s^2} \int [f^{(s)}(x)]^2\, dx \left(\int u^s K(u)\, du \right)^2 \tag{1.8}$$

as $n \to \infty$. We shall present the proof of (1.8) in the case when $K(x) \geq 0$. The general case follows from the representation $K(x) = K^+(x) + K^-(x)$ where $K^+(x)$ and $K^-(x)$ are respectively the positive and negative parts of $K(x)$.

In accordance with Hölder's inequality and Fubini's theorem we have

$$\limsup_n A_n \leq \int [f^{(s)}(x)]^2\, dx \left(\frac{1}{s} \int u^s K(u)\, du \right)^2. \tag{1.9}$$

On the other hand, it follows from the Lebesgue theorem that

$$\int\int_0^1 (1-t)^{s-1} f^{(s)}\left(x + t\frac{u}{a_n}\right) u^s K(u)\, du\, dt \to f^{(s)}(x) \frac{1}{s} \int u^s K(u)\, du$$

as $n \to \infty$ and hence an application of Fatou's lemma yields:

$$\int [f^{(s)}(x)]^2\, dx \left(\frac{1}{s} \int u^s K(u)\, du \right)^2 \leq \liminf_n A_n. \tag{1.10}$$

Combining the inequalities (1.9) and (1.10) we arrive at (1.8). The theorem is thus proved.

COROLLARY. For $f(x) \in W_2$ and $K(x) = 1/2$ for $|x| < 1$ and $K(x) = 0$ for $|x| \geq 1$ (1.5) is reduced to Rosenblatt's result [4].

We shall now determine the optimal value $a_n = a_n^0$ which minimizes the asymptotic expression (as $n \to \infty$) of the m.i.s.e. $U_n(a_n)$. From (1.5) we obtain

$$a_n^0 = [2sn\alpha^2 \int [f^{(s)}(x)]^2\, dx ((s!)^2 \int K^2(x)\, dx)^{-1}]^{1/(2s+1)} =$$
$$= \Theta n^{\gamma}, \tag{1.11}$$

where

$$\theta^{2s+1} = 2s\alpha^2 \int [f^{(s)}(x)]^2\, dx((s!)^2 \int K^2(x)\, dx)^{-1}, \quad \gamma = \frac{1}{2s+1}.$$

Substituting the optimal value of a_n^0 obtained into the r.h.s. of expression (1.5) we have as $n \to \infty$

$$U(a_n^0) = \Theta(s, f, K) n^{-2s/(1+2s)} + o(n^{-2s/(1+2s)}),$$

where

$$\Theta(s, f, K) = (1 + 2s)\{((2s)^{-1} \int K^2(u)\, du)^{2s} (s!)^{-2} \times \\ \times \alpha^2 \int [f^{(s)}(x)]^2\, dx\}^{1/(1+2s)}.$$

Thus the family of estimators $f_n(x, a_n^0)$ is integrally consistent of the order $N(n) = n^{2s/(1+2s)}$; this means that $N(n)\, U(a_n^0)$ approaches a finite non-zero limit as $n \to \infty$.

The following lemma is proved in the same manner as relation (1.8).

LEMMA 1.1. If $f(x) \in W_s$ and $K(x) \in H_s$ and if, moreover $K(x)$ possesses continuous derivatives up to the s-th order inclusively, and also

$$K^{(i)}(x) \to 0, \quad i = \overline{1, s-1}, \quad \text{as} \quad x \to \pm\infty \quad \text{and}$$

$$\int x^s |K^{(s)}(x)|\, dx < \infty, \quad \text{then as} \quad n \to \infty$$

$$\int (Ef_n^{(s)}(x, a_n))^2\, dx \to \int (f^{(s)}(x))^2\, dx. \tag{1.12}$$

<u>Remark 1</u>. The estimators $f_n^{(i)}(x, a_n)$ of the derivatives $f^{(i)}(x)$, $i = 1, 2, \ldots$ will be discussed in more detail in Section 2 of Chapter 1 and in Section 5 of Chapter 4.

<u>Remark 2</u>. Let a sample X_j, $j = \overline{1, n}$ for a population distributed with density $f(\vec{x})$, $\vec{x} = (x_1, x_2, \ldots, x_p)$ be given. As an estimator of $f(\vec{x})$ we shall consider the statistic

$$f_n(\vec{x}, a_n) = \frac{a_n^p}{n} \sum_{i=1}^{n} K(a_n(\vec{x} - X_i)),$$

where

$$K(\vec{x}) = \prod_{j=1}^{n} K_j(x_j) \quad \text{and} \quad K_j(x_j) \in H_s, \quad j = \overline{1, p}.$$

Let $U(a_n)$ denote the m.i.s.e. of the estimator $f_n(x, a_n)$ in this case as well.

The following generalization of Theorem 1.1 is proved analogously.

THEOREM 1.2. If $f(x)$ possesses all the partial derivatives up to the s-th order inclusively and these derivatives are continuous, bounded and belong to $L_2(R_p)$ and $K_j \in H_s$, $j = \overline{1, p}$, then as $n \to \infty$

$$U(a_n) = \frac{a_n^p}{n} \int K^2(\vec{x})\, d\vec{x} + \frac{a_n^{-2s}}{(s!)^2} \int \left[\sum_{j=1}^{p} \alpha_j \frac{\partial^{(s)} f(\vec{x})}{\partial x_j^s} \right]^2 d\vec{x} + $$
$$+ O\left(\frac{a_n^p}{n} + a_n^{-2s} \right),$$

where

$$\alpha_j = \int u^s K_j(u)\, du.$$

The optimal sequence $\{a_n^0\}$ for which $U(a_n)$ possesses the best order of smallness is

$$a_n^0 = \Theta_1(f) n^{1/(2s+p)},$$

and for the optimal a_n^0

$$U(a_n^0) = \Theta_2(f) n^{-2s/(p+2s)} + o(n^{-2s/(p+2s)}),$$

where Θ_1 and Θ_2 are functionals in $f(\vec{x})$ (as in the univariate case Θ_1 and Θ_2 can be written explicitly).

Below the following lemma will be used.

LEMMA 1.2 ([14]). Let the random variables η_n and ξ_n possess absolute moments of the m-th order and let with probability 1, $\eta_n \geq d_n > 0$, $\xi_n \geq 2\delta > 0$ and $d_n \to 0$ as $n \to \infty$. If $E|\eta_n - \xi_n|^m = O(a_n^m)$ where $a_n^k = O(d_n)$ for some $k > 0$ then $E|\eta_n^\alpha - \xi_n^\alpha|^m = O(a_n^m)$ with $\gamma_1^0 < \alpha \leq 1$.

3. We now proceed to a solution of the second problem posed in the beginning of this subsection.

Assume that $f(x)$ and $K(x)$ satisfy the conditions of Lemma 1.1. Let $\{t_n\}$ be a sequence of positive numbers such that $t_n \to \infty$ as $n \to \infty$ and moreover, let $t_n = O(n^\beta)$, $\beta = 1/2(2s + 1)^2$. Furthermore, let $\{b_n\}$ be a sequence of positive numbers which converge to zero while satisfying the condition $nb_n \geq C > 0$ ($C, C_1, C_2, \ldots$ denoted here and below by positive constants).

We shall firstly estimate the parameter Θ appearing in (1.11) based on the observations $X_1, X_2, \ldots, X_n$. Denote $f_n^{(s)}(x) = f_n^{(s)}(x; t_n)$, $\mu_n^{(s)}(x) = Ef_n^{(s)}(x; t_n)$ and choose the expression

$$\hat{\Theta}_n^{2s+1} = \ell(K, s)[\int [f_n^{(s)}(x)]^2\, dx + b_n], \tag{1.13}$$

where

$$\ell(K, s) = 2s(\int x^s K(x)\, dx)^2((s!)^2 \int K^2(x)\, dx)^{-1}.$$

to be the estimator of parameter Θ.
Set

$$\Theta_n^{2s+1} = \ell(K, s)[\int (\mu_n^{(s)}(x))^2\, dx + b_n],$$

$$\hat{a}_n = \hat{\Theta}_n n^{\gamma}, \quad \sigma_n = \Theta_n n^{\gamma}, \quad \tau_n^2 = \frac{t_n^{2(2s+1)}}{n} = O(n^{-2s\gamma}).$$

Let

$$K^*(u) = \int K^{(s)}(v) K^{(s)}(u - v)\, dv,$$
$$T_n^*(u) = t_n \int K^*(t_n(u - v)) f(v)\, dv. \tag{1.14}$$

It is easy to verify that

$$ET_n^*(X) = t_n^{-2s} \int [\mu_n^{(s)}(z)]^2\, dz, \tag{1.15}$$

where X is a random variable distributed with the density f(x).

LEMMA 1.3. If f(x) and K(x) satisfy the conditions of Lemma 1.1 then

$$E|\hat{\Theta}_n - \Theta_n|^m = O(\tau_n^m), \tag{1.16}$$

where $m > 0$ is an integer.

Proof. Utilizing (1.14) and (1.15) we have

$$E[\int (f_n^{(s)}(x))^2\, dx - \int (\mu_n^{(s)}(x))^2\, dx]^{2m} =$$
$$= t_n^{4sm} E[n^{-2} \sum_{j=1}^{n} \sum_{i=1}^{m} t_n K^*(t_n(X_j - X_i)) - ET_n^*(X)]^{2m}.$$

Whence in view of the inequality

$$\left| \sum_{k=1}^{m} z_k \right|^p \leq m^{p-1} \sum_{k=1}^{m} |z_k|^p,$$

$p \geq 0$ is an integer, we obtain

$$E[\int (f_n^{(s)}(x))^2\, dx - \int (\mu_n^{(s)}(x))^2\, dx]^{2m} \leq c_1 t_n^{4ms}(E_n^{(1)} + E_n^{(2)}), \tag{1.17}$$

where

$$E_n^{(1)} = E[n^{-2} \sum_{j=1}^{m} \sum_{i=1}^{n} t_n K^*(t_n(X_j - X_i)) - n^{-1} \sum_{j=1}^{n} T_n^*(X_j)]^{2m},$$

$$E_n^{(2)} = E[n^{-1} \sum_{j=1}^{n} (T_n^*(X_j) - ET_n^*(X_j))]^{2m}.$$

We bound each one of the expressions $E_n^{(1)}$ and $E_n^{(2)}$ separately.

We have the inequalities

$$|K^*(u)| \leq C_2, \quad |T_n^*(u)| \leq C_2 t_n; \tag{1.18}$$

and, hence,

$$E_n^{(1)} \leq C_3 \left(\frac{t_n}{n}\right)^{2m} + E\left| n^{-1} \sum_{j=1}^{n} \left[n^{-1} \sum_{i=1(\neq j)} t_n K^*(t_n(X_j - X_i)) - \right.\right.$$

$$\left.\left. - T_n^*(X_j)\right]\right|^{2m} \leq C_3 \left(\frac{t_n}{n}\right)^{2m} + n^{-1} E \sum_{j=1}^{n} \left|\left[\sum_{\substack{1 \leq i \leq n \\ i \neq j}} t_n K^* \times \right.\right.$$

$$\left.\left. \times (t_n(X_j - X_i)) - T_n^*(X_j)\right]\right|^{2m} = C_3 \left(\frac{t_n}{n}\right)^{2m} +$$

$$+ E\left[n^{-1} \sum_{j=2}^{n} t_n K^*(t_n(X_j - X_1)) - T_n^*(X_1)\right]^{2m} \leq C_4 \left(\frac{t_n}{n}\right)^{2m} +$$

$$+ E\left\{ E\left(\frac{1}{n-1} \sum_{i=2}^{n} t_n K^*(t_n(X_i - X_1)) - T_n^*(X_1)\right)^{2m} / X_1 \right\} =$$

$$= C_4 \left(\frac{t_n}{n}\right)^{2m} + E_n^{(3)}. \tag{1.19}$$

Thus the bound on $E_n^{(1)}$ is reduced to a bound on $E_n^{(3)}$. We shall utilize von Bahr's results [29] dealing with the convergence of moments in the central limit theorem.

Denote

$$\sigma_n^2(X_1) = D[t_n K^*(t_n(X - X_1))/X_1],$$

$$V_n(X_i, X_1) = [t_n K^*(t_n(X_i - X_1)) - T_n^*(X_1)]\sigma_n^{-1}(X_1), \quad i = \overline{2, n}.$$

We have

$$E_n^{(3)} = E\left\{ \left(\frac{\sigma_n(X_1)}{\sqrt{n-1}}\right)^{2m} E\left(\left(\sum_{i=2}^{n} V_n(X_i, X_1) \frac{1}{\sqrt{n-1}}\right)^{2m} / X_1\right) \right\}. \tag{1.20}$$

For a given X_1, $V_n(X_i, X_1)$, $i = \overline{2, n}$ are independent identically distributed random variables such that

$$E(V_n(X_i, X_1)/X_1) = 0, \quad D(V_n(X_i, X_1)/X_1) = 1.$$

Therefore, in view of Theorem 1 in von Bahr's paper [29] we obtain

$$E\left(\left(\frac{1}{\sqrt{n-1}}\sum_{i=2}^{n} V_n(X_i, X_1)\right)^{2m} / X_1\right) = \int u^{2m}\phi(u)\, du +$$

$$+ \sum_{\nu=1}^{2m-2} (n-1)^{-\nu/2} \int u^{2m} p_{n\nu}(-\phi)(u)\, du, \tag{1.21}$$

where

$$p_{n\nu}(-\phi)(u) = \Pi_{3\nu,n}(u)e^{-u^2/2}, \quad \phi(u) = (\sqrt{2\pi})^{-1}e^{-(u^2/2)},$$

and $\Pi_{3\nu,n}(u)$ is a polynomial in u of degree 3ν whose coefficients depend only on the moments

$$\alpha_n^{(r)}(X_1) = E[(V_n(X, X_1)^r/X_1], \quad r = 2, \ldots, \nu + 2.$$

More precisely, from Petrov's theorem [30] we have

$$p_{\nu n}(-\phi)(u) = \phi(u)\sum \frac{1}{k_1!\ldots k_\nu!}\left(\frac{\gamma_{n_3}}{3!}\right)^{k_1} \cdots \left(\frac{\gamma_{n\nu+2}}{(\nu+2)!}\right)^{k_\nu} \times$$

$$\times H_{3k_{1+\ldots+(\nu+2)}k_\nu}(u), \tag{1.22}$$

where the summation is carried out over all the integer-valued non-negative solutions of the equation $k_1 + 2k_2 + \ldots + \nu k = \nu$, γ_{nj} is the j-th cumulant of the random variable $V_n(X, X_1)$ for a given X_1 (X is one of the random variables $X_2, \ldots, X_n$) and $H_N(x)$ is the Chebyshev-Hermite polynomial of degree N.

Since

$$|\gamma_{nj}| \leq j^j E(|V_n(X, X_1)|^j/X_1),$$

in view of (1.18) we have

$$|\gamma_{nj}| \leq C_2^{j-2} j^j t_n^{j-2} \sigma_n^{2-j}(X_1), \tag{1.23}$$

and hence it follows from here and from (1.22) that

$$\left|\int u^{2m} p_{n\nu}(-\phi)(u)\, du\right| \leq C_3 \frac{t_n^\nu}{[\sigma_n^2(X_1)]^{\nu/2}}, \tag{1.24}$$

where $C_3 = C_3(\nu, m)$ is a constant which depends on ν and m.

In view of (1.21) and (1.24) and taking $\sigma_n^2(X_1) \leq C_4 t_n^2$ into account we obtain from (1.20) that:

$$E_n^{(3)} = O\left(\left(\frac{t_n^2}{n}\right)\right)^m . \tag{1.25}$$

Thus from (1.19) and (1.25) we have

$$E_n^{(1)} = O\left(\left(\frac{t_n^2}{n}\right)\right)^m . \tag{1.26}$$

Repeating the same arguments as in the course of estimating $E_n^{(3)}$ we conclude that

$$E_n^{(2)} = O\left(\left(\frac{t_n^2}{n}\right)\right)^m . \tag{1.27}$$

Next, substituting (1.26) and (1.27) into (1.17) we obtain:

$$E\left[\int (f_n^{(s)}(x, t_n))^2\, dx - \int (\mu_n^{(s)}(x))^2\, dx\right]^{2m} = O(\tau_n^{2m}). \tag{1.28}$$

And hence recalling the definition of $\hat{\theta}_n^{2s+1}$ and θ_n^{2s+1}, from (1.28) we have

$$E\left|\hat{\theta}_n^{2s+1} - \theta_n^{2s+1}\right|^{2m} = O(\tau_n^{2m}). \tag{1.29}$$

From here we have in particular that

$$E\left|\hat{\theta}_n^{2s+1} - \theta_n^{2s+1}\right|^{m} = O(\tau_n^{m}). \tag{1.30}$$

Finally since $\hat{\theta}_n^{2s+1} \geq \ell(K, s) b_n$, $\tau_n^4 = O(b_n)$ and in view of Lemma 1.1 $\theta_n^{2s+1} \geq \theta^{2s+1}/2$ for $n \geq N$, in accordance with (1.30) and Lemma 1.2 we arrive at (1.16). The lemma is thus proved.

COROLLARY. It follows from Lemma 1.1 and (1.16) that $\hat{\theta}_n$ is a consistent estimator of θ.

We now prove the basic theorem which will allow us to solve the problem posed above.

THEOREM 1.3. Let $f(x) \in W_s$, $K(x) \in H_s$ and moreover, let $K(x)$ possess continuous derivatives of the order s inclusively and let $K^{(i)}(x) \to 0$ as $x \to \pm\infty$, $i = \overline{1, s-1}$ and $\int x^s |K^{(s)}(x)|\, dx < \infty$. If the function $K_1(x) = K(x) + xK^{(1)}(x)$ admits on the interval $(0, \infty)$ a non-increasing and integrable majorant $K_0(x)$, $K_0(x) = K_0(-x)$ then as $n \to \infty$

$$U(\hat{a}_n) \sim U(a_n^0) \tag{1.31}$$

(here and below the relation $\alpha_n \sim \beta_n$ between variable quantities α_n and β_n means that $\alpha_n/\beta_n \to 1$ as $n \to \infty$).

<u>Proof</u>. In view of (1.5) and (1.12) $\Theta_n \to \Theta$, $U_n(\sigma_n) \sim U_n(a_n^0)$ as $n \to \infty$. Moreover, one easily observes that to prove (1.31) it is sufficient to show that

$$E \int (f_n(x, \hat{a}_n) - f_n(x, \sigma_n))^2 \, dx = O(n^{-2s\gamma}). \tag{1.32}$$

We utilize the formula of finite differences and the Cauchy-Schwarz inequality:

$$E \int [f_n(x, \hat{a}_n) - f_n(x, \sigma_n)]^2 \, dx \le$$

$$\le E^{1/2} \left[\int \left(\frac{\partial f_n(x, t)}{\partial t} \right)^2_{t=\xi_n} dx \right]^2 E^{1/2} (\hat{\Theta}_n - \Theta_n)^4 n^{2\gamma},$$

where the random variable ξ_n is located between $\hat{a}_n$ and σ_n:

$$|\sigma_n - \xi_n| < |\hat{a}_n - \sigma_n|.$$

However, in view of (1.16)

$$E^{1/2} (\hat{\Theta}_n - \Theta_n)^4 = O(\tau_n^2) = O(n^{-2\gamma s}).$$

Clearly, therefore, it remains only to verify that

$$n^{4\gamma} E \left[\int \left(\frac{\partial f_n(x, t)}{\partial t} \right)^2_{t=\xi_n} dx \right]^2 = O(1). \tag{1.33}$$

Let A_n be the event that $\hat{\Theta}_n > \frac{\Theta}{2}$ and I_{A_n} be the indicator of A_n. Next, let $\lambda_n = \frac{1}{2} \Theta n^{\gamma}$. We have

$$n^{4\gamma} E \left[\int \left(\frac{\partial f_n(x, t)}{\partial t} \right)^2_{t=\xi_n} dx \right]^2 =$$

$$= n^{4\gamma} E \left[\int \left(\frac{1}{n} \sum_{i=1}^{n} K_1(\xi_n(x - X_i)) \right)^2 \right]^2 = n^{4\gamma} (B_n^{(1)} + B_n^{(2)}),$$

where

$$B_n^{(1)} = E \left[\int \left(\frac{1}{n} \sum_{j=1}^{n} K_1(\xi_n(x - X_j)) \right)^2 dx \, I_{A_n} \right]^2,$$

$$B_n^{(2)} = E \left[\int \left(\frac{1}{n} \sum_{i=1}^{n} K_1(\xi_n(x - X_i)) \right)^2 dx \, I_{\bar{A}_n} \right]^2$$

For n sufficiently large, $\sigma_n \geq \frac{1}{2}\theta n^\gamma$ and $\hat{a}_n \geq \frac{1}{2}\theta n^\gamma$ on the points ω of the set A_n and hence $\xi_n \geq \lambda_n = \frac{1}{2}\theta n^\gamma$ for $\omega \in A_n$. Next, taking into account that $K_1(x)$ is majorized by $K_0(x)$ we obtain:

$$n^{4\gamma} B_n^{(1)} \leq C_5 E \left[\int \left(\frac{\lambda_n}{n} \sum_{i=1}^{n} K_0(\lambda_n(x - X_i)) \right)^2 dx \right]^2.$$

Based on the method used in the proof of (1.28) and on Lemma (1.1) we conclude:

$$E \left[\int \left(\frac{\lambda_n}{n} \sum_{i=1}^{n} K_0(\lambda_n(x - X_i)) \right)^2 dx \right]^2 = O(1).$$

Hence

$$n^{4\gamma} B_n^{(1)} = O(1). \tag{1.34}$$

Now set

$$K_0^*(x) = \int K_0(v) K_0(x - v)\, dv.$$

Then

$$B_n^{(2)} \leq E \left[\frac{n^{-2}}{\xi_n} \sum_{i=1}^{n} \sum_{j=1}^{n} K_0^*(\xi_n(X_i - X_j) I_{\bar{A}_n} \right]^2 \leq C_6 E \left(\frac{1}{\xi_n} I_{\bar{A}_n} \right)^2.$$

In view of (1.3) and the fact that $nb_n \geq C > 0$ with probability 1 we have

$$\hat{\tau}_n = \hat{\theta}_n n^\gamma \geq [\ell(K, s)]^\gamma (nb_n)^\gamma \geq [C\ell(K, s)]^\gamma = C_7 \neq 0,$$

and also $\sigma_n \geq C_7$. This implies that $\xi_n \geq C_7$. Consequently,

$$n^{4\gamma} B_n^{(2)} \leq C_8 n^{4\gamma} P(\bar{A}_n).$$

Next, taking (1.16) and Chebyshev's inequality into account we arrive at:

$$P(\bar{A}_n) = O(n^{-4s\gamma}).$$

Hence

$$n^{4\gamma} B_n^{(2)} = O(1), \tag{1.35}$$

relations (1.34) and (1.35) imply (1.33) and the theorem is thus proved.

Remark 3. It follows from (1.32) and the expansion of $U_n(a_n^0)$ that

$$U(\hat{a}_n) = U(a_n^0) + O(n^{-2s/(1+2s)}).$$

Remark 4. The quantity $E\|f_n - f\|_p^p$, $p \geq 2$, can be chosen as the measure of error of the estimator $f_n(x, a_n)$; here p is an even number ($\|\cdot\|_p$) is the norm in the space $L_p(-\infty, \infty)$).

THEOREM 1.4. Let $f \in W_s$ and $K \in H_s$ and moreover, let $f^{(s)} \in L_p$, $f(x) \in L_{(p-m)/2}$, $m = 0, 1, \ldots, p-2$. Then choosing $a_n = C_9 n^{1/(1+2s)}$,

$$\limsup_n n^{sp/(1+2s)} E \, \| f_n - f \|_p^p \leq C(f, K, s, p) < \infty,$$

where $C(f, K, s, p)$ is a constant which depends on f, K, s and p.

Proof. We have

$$E \| f_n - f \|_p^p \leq 2^{p-1} (E \| f_n - Ef_n \|_p^p + \| f - Ef_n \|_p^p).$$

We shall represent $\| f - Ef_n \|_p^p$ in a form analogous to (1.7) and then applying the generalized Minkowski inequality [31] we arrive at

$$\| f - Ef_n \|_p^p \leq \frac{a_n^{-ps}}{(s!)^p} \| f^{(s)} \|_p^p \left(\int u^s K(u) \, du \right)^p = C_{10} a_n^{-ps}. \quad (1.36)$$

We bound $E \| f_n - Ef_n \|_p^p$. The expression $f_n(x, a_n) - Ef_n(x, a_n)$ is the arithmetic mean of independent random variables $\xi_1, \ldots, \xi_n$ with mean 0, each one of which is distributed as $\xi_1 = \xi_1(x) = a_n[K(a_n(x - X_1)) - EK(a_n(x - X_1))]$. Denote $V(x) = D\xi_1(x)$ and $\eta_i = \xi_i/\sqrt{V}$. For the normalized sum

$$[f_n(x, a_n) - Ef_n(x, a_n)](Df_n(x, a_n))^{-1/2} = \frac{1}{\sqrt{n}} \sum_{i=1}^{n} \eta_i,$$

formulas analogous to (1.21)-(1.24) are valid. Based on these formulas we obtain

$$\begin{aligned} E \| f_n - Ef_n \|_p^p &= n^{-p/2} \int V^{p/2}(x) \left[\int u^p \phi(u) \, du + \right. \\ &\left. + \sum_{\nu=1}^{p-2} n^{-\nu/2} \int u^p P_{n\nu}(-\phi)(u) \, du \right] dx \leq C_{11} n^{-p/2} \times \\ &\times \int \left(\sum_{\nu=0}^{p-2} n^{-\nu/2} V^{(p-\nu)/2}(x) a_n^{\nu} \right) dx \leq \\ &\leq C_{11} \sum_{\nu=0}^{p-2} \left(\frac{a_n}{n} \right)^{(p+\nu)/2} \int V_0^{(p-\nu)/2}(x) \, dx, \end{aligned} \quad (1.37)$$

where $V(x) \leq V_0(x) = a_n \int K^2(a_n(x - u)) f(u) \, du$. It follows from the generalized Minkowski inequality and the condition of the theorem that the integrals apprearing on the r.h.s. of (1.37) are finite. Therefore

$$E\,||f_n - Ef_n||_p^p \le C_{12}\left(\frac{a_n}{n}\right)^{p/2}. \tag{1.38}$$

Hence

$$E\,||f_n - f\,||_p^p \le C_{13}\left(\frac{a_n}{n}\right)^{p/2} + C_{14}a_n^{-ps}.$$

Evidently the best order of smallness on the r.h.s. of the last inequality is attained by setting $a_n = C_9 n^{1/(1+2s)}$. For this choice of a_n we have $E\,||f_n - f\,||_p^p \le C_{15}n^{-(ps)/(1+2s)}$. The theorem is thus proved.

2. The Mean Square Error of Nonparametric Kernel-Type Density Estimators

1. Let $X = (X^{(1)}, X^{(2)}, \ldots, X^{(p)})$ be a random variable with values in a Euclidean p-dimensional space R_p, and the function $f(\vec{x})$, $\vec{x} = (x_1, \ldots, x_p)$ be the unknown probability density of the random variable X. Next, let a sample $X_j = (X_j^{(1)}, X_j^{(2)}, \ldots, X_j^{(p)})$, $j = \overline{1, n}$ from the population X be given. As an estimator of the unknown density $f(\vec{x})$ based on the sample X we shall consider an estimator which generalizes the estimator presented in Remark 2 of Section 1 of Chapter 1 of the form ([32], [33], [34]):

$$f_n(\vec{x}, \vec{a}_n) = n^{-1}\sum_{j=1}^{n}\prod_{i=1}^{p} a_i(n)K_i(a_i(n)(x_i - X_j^{(i)})), \tag{2.1}$$

where

$$K_j(u) \in H_s, \quad j = \overline{1, p}, \quad \vec{a}_n = (a_1(n), \ldots, a_p(n)),$$

and $\{a_j(n)\}$, $j = \overline{1, p}$ is a sequence of positive numbers such that $\lim_{n\to\infty} a_j(n) = \infty$, $j = \overline{1, p}$ and

$$\lim_{n\to\infty} n^{-1}\prod_{i=1}^{p} a_i(n) = 0.$$

It is known (c.f. e.g., [32], [33]) that certain asymptotic properties of the estimator (2.1) for a fixed $\vec{x}$ as $n \to \infty$ depend on the smoothness of f(x) in the neighborhood of the point $\vec{x}$, $\vec{x} \in R_p$. For example, in the case $a_1(n) = \ldots = a_p(n) = a(n)$ the coefficient C of the optimal $a(n) = a^0(n)$ ($a^0(n) = Cn^{\gamma}$ where γ is a known number defined below) obtained by means of minimizing the asymptotic expression for the mean square error (m.s.e.) $E(f_n(\vec{x}, \vec{a}_n) - f(\vec{x}))^2$, depends on $f(\vec{x})$ and on its partial derivatives. In general, it is difficult to use such an $a^0(n)$ in in practice. A problem thus naturally arises as to whether one can replace the sequence $\{a_n\}$ in the estimator (2.1) by some known sequence of random variables which depends on the sample values X_j, $j = \overline{1, n}$ in such a manner that the m.s.e. of the estimator obtained will be equivalent (as $n \to \infty$) to the m.s.e. of the estimator (2.1) for an optimal $a^0(n)$.

The aim of Section 2 is to solve this problem, to obtain an asymptotic expansion of the m.s.e. and to study certain properties of estimators of partial derivatives of the probability density.

2. We shall assume that $f(\vec{x})$ is bounded on R_p and is continuous - together with its partial derivatives up to the s-th order inclusively - in a certain neighborhood of point $\vec{x}$, and moreover $f(\vec{x}) \neq 0$. We shall denote these assumptions by G_s.

We introduce the following notation: $\|\cdot\|$ is the length of the vector $\vec{x} = (x_1, \ldots, x_p)$, $b(f_n) = Ef_n(\vec{x}, \vec{a}_n) - f(x)$, $\alpha_j = \int x^s K_j(x)\, dx$, C_j, $j = 1, 2, \ldots$, are positive numbers which may or may not depend on $\vec{x}$.

LEMMA 2.1. Let $f(\vec{x})$ satisfy the conditions G_s and $K_j(u) \in H_s$, $u \in R_1$, $j = \overline{1, p}$. If $\lim_{n\to\infty} a_i(n)/a_j(n) = a_{ij} > 0$, $i \neq j$, $i, j = 1, 2, \ldots, p$, then as $n \to \infty$

$$b(f_n) \sim \frac{\|\vec{a}_n\|^{-s}}{s!} \sum_{i=1}^{p} \|\vec{b}_i\|^s \alpha_i \frac{\partial^s f(\vec{x})}{\partial x_i^s}, \tag{2.2}$$

where

$$\vec{b}_i = (a_{1i}, \ldots, a_{pi}), \quad a_{ii} = 1, \quad i = \overline{1, p}.$$

Proof. Let $\delta > 0$. Since $f(\vec{x}) \in G_s$ and $K_j(u) \in H_s$ one can write*

$$\left| \|\vec{a}_n\|^s b(f_n) - \frac{1}{s!} \sum_{i=1}^{p} \|\vec{b}_i\|^s \alpha_i \frac{\partial^s f(\vec{x})}{\partial x_i^s} \right| \leq \|\vec{a}_n\|^s \left| b(f_n) - \frac{1}{s!} \sum_{i=1}^{p} \frac{\alpha_i}{a_i^s} \frac{\partial^s f(\vec{x})}{\partial x_i^s} \right| + \left| \frac{1}{s!} \sum_{i=1}^{p} \alpha_i \frac{\|\vec{a}_n\|}{a_i^s} \times \frac{\partial^s f(\vec{x})}{\partial x_i^s} - \frac{1}{s!} \sum_{i=1}^{p} \|\vec{b}_i\|^s \alpha_i \frac{\partial^s f(\vec{x})}{\partial x_i^s} \right| =$$

*In place of $a_i(n)$ we use the notation a_i.

$$= \|\vec{a}_n\|^s \left| \int \left[f\left(x_1 + \frac{u_1}{a_i}, \ldots, x_p + \frac{u_p}{a_p}\right) - \sum_{i=1}^{s} \frac{1}{i!} \times \right.\right.$$

$$\left. \times \left(\frac{\partial}{\partial x_i}\frac{u_1}{a_1} + \ldots + \frac{\partial}{\partial x_p}\frac{u_p}{a_p}\right)^{(i)} f(\vec{x}) \right] \prod_{j=1}^{p} K_j(u_j)\, du_j \Bigg| +$$

$$+ \left| \frac{1}{s!} \sum_{i=1}^{p} \alpha_i \frac{\|\vec{a}_n\|^s}{a_i^s} \frac{\partial^s f(\vec{x})}{\partial x_i^s} - \frac{1}{s!} \sum_{i=1}^{p} \|\vec{b}_i\|^s \alpha_i \times \right.$$

$$\left. \times \frac{\partial^s f(\vec{x})}{\partial x_i^s} \right| \leq \|\vec{a}_n\|^s \int_{\|\vec{z}_n\| \leq \delta} |R_s(\vec{x}, \vec{z}_n)|\, \|\vec{z}_n\|^s \times$$

$$\times \prod_{j=1}^{p} |K_j(u_j)|\, du_j + \|a_n\|^s \int_{\|\vec{z}_n\| > \delta} \Bigg| \times$$

$$\times f\left(x_1 + \frac{u_1}{a_1}, \ldots, x_p + \frac{u_p}{a_p}\right) - \sum_{i=0}^{s} \frac{1}{i!} \left(\frac{\partial}{\partial x_1}\frac{u_1}{a_i} + \ldots + \right.$$

$$\left. + \frac{\partial}{\partial x_p}\frac{u_p}{a_p}\right)^{(i)} f(\vec{x}) \Bigg| \prod_{j=1}^{p} \Big| K_j(u_j) \Big| du_j + \left| \frac{1}{s!} \sum_{i=1}^{p} \alpha_i \times \right.$$

$$\left. \times \frac{\|\vec{a}_n\|^s}{a_i^s} \frac{\partial^s f(\vec{x})}{\partial x_i^s} - \frac{1}{s!} \sum_{i=1}^{p} \|\vec{b}_i\|^s \alpha_i \frac{\partial^s f(x)}{\partial x_i^s} \right| =$$

$$= I_n^{(1)} + I_n^{(2)} + I_n^{(3)},$$

where

$$R_s(\vec{x}, \vec{y}) \to 0, \quad \text{as} \quad \|\vec{y}\| \to 0 \quad \text{and} \quad \vec{z}_n = \left(\frac{u_1}{a_1}, \ldots, \frac{u_p}{a_p}\right).$$

Set

$$M = \max_{\vec{x}} \left\{ \left| \frac{\partial^m f(\vec{x})}{\partial x_1^{k_1} \ldots \partial x_p^{k_p}} \right|, \quad k_1 + \ldots + k_p = m, \quad 0 \leq m \leq s \right\}.$$

Since

$$\left| \sum_{j=1}^{p} x_j \right|^m \leq p^{m-1} \sum_{j=1}^{p} |x_j|^m,$$

we have

$$I_n^{(1)} \leq \max_{\|\vec{z}\| \leq \delta} |R_s(\vec{x}, \vec{z})| \int_{\|\vec{z}_n\| \leq \delta} \|\vec{a}_n\|^s \|\vec{z}_n\|^s \prod_{j=1}^{p} \times$$

$$\times |K_j(u_j)| \, du_j \leq p^{s-2} \max_{\|\vec{z}\| \leq \delta} |R_s(\vec{x}, \vec{z})| \sum_{i=1}^{p} \times$$

$$\times \frac{a_1^s + \ldots + a_p^s}{a_i^s} \int u_i^s \prod_{j=1}^{p} |K_j(u_j)| \, du_j,$$

$$I_n^{(2)} \leq 2M \|\vec{a}_n\|^s \int_{\|\vec{z}_n\| \geq \delta} \sum_{m=0}^{s} \left(\frac{|u_1|}{a_1} + \ldots + \frac{|u_p|}{a_p} \right)^m \prod_{j=1}^{p} \times$$

$$\times |K_j(u_j)| \, du_j \leq 2M \|\vec{a}_n\|^s p^s \int_{\|\vec{z}_n\| \geq \delta} \sum_{m=0}^{s} \|\vec{z}_n\|^m \prod_{j=1}^{p} \times$$

$$\times |K_j(u_j)| \, du_j \leq 2M\delta^{-s} p^s \|\vec{a}_n\| \int_{\|z_n\| \geq \delta} \|\vec{z}_n\|^s \sum_{m=0}^{s} \delta^m \times$$

$$\times \prod_{j=1}^{p} |K_j(u_j)| \, du_j \leq \frac{2M\delta^{-s} p^{2s-2}}{1 - \delta} \sum_{i=1}^{p} \frac{a_1^s + \ldots + a_p^s}{a_i^s} \times$$

$$\times \int_{\|\vec{y}\| \geq \delta \min\limits_{1 \leq i \leq p} a_i} y_i^s \prod_{j=1}^{p} |K_j(y_j)| \, dy_j.$$

Hence

$$\left| \|\vec{a}_n\|^s b(f_n) - \frac{1}{\vec{s}!} \sum_{i=1}^{p} \|\vec{b}_i\|^s \alpha_i \frac{\partial^s f(\vec{x})}{\partial x_i^s} \right| \leq p^{s-2} \max_{\|\vec{z}\| \leq \delta} \times$$

$$\times |R_s(\vec{x}, \vec{z})| \sum_{i=1}^{p} \frac{a_1^s + \ldots + a_p^s}{a_i^s} \int u_i^s \prod_{j=1}^{p} |K_j(u_j)| \, du_j +$$

$$+ \frac{2M\delta^{-s} p^{2s-2}}{1 - \delta} \sum_{i=1}^{p} \frac{a_1^s + \ldots + a_p^s}{a_i^s} \int_{\|\vec{u}\| \geq \min\limits_{1 \leq j \leq p} a_j} u_i^s \prod_{j=1}^{p} |K_j(u_j)| \times$$

$$\times du_j + \left| \frac{1}{s!} \Sigma \alpha_i \frac{\|\vec{a}_n\|^s}{a_i^s} \frac{\partial^s f(\vec{x})}{\partial x_i^s} - \frac{1}{s!} \sum_{i=1}^{p} \alpha_i \|\vec{b}_i\|^s \frac{\partial^s f(x)}{\partial x_i^s} \right|.$$

Taking the properties of $f(\vec{x})$ and $K_j(u)$, $u \in R_1$, into account, we verify that the r.h.s. of the last inequality can be made arbitrarily small by choosing first a small δ and then an n sufficiently large. The lemma is thus proved.

COROLLARY 1. Theorem 4.2 of paper [33] follows from Lemma 2.1 by setting $f(\vec{x}) \in G_2$ and $K_j(u) \in H_2$, $j = \overline{1, p}$.

Below we shall assume that $a_1(n) = a_2(n) = \ldots = a_p(n) = a_n$.

LEMMA 2.2. If $K_j(x) \in H_0$ and $a_n^p/n \to 0$ as $n \to \infty$, then

$$Df_n(\vec{x}, a_n) \sim \frac{a_n^p}{n} f(\vec{x}) \prod_{i=1}^{p} \int K_i^2(x)\,dx, \tag{2.3}$$

at each point $\vec{x} \in R_p$, where $f(\vec{x})$ is continuous.

The proof of this lemma is analogous to the proof in the one-dimensional case [5].

THEOREM 2.1. If $f(\vec{x}) \in G_s$ and $K_j(u) \in H_s$, $u \in R_1$, $j = \overline{1, p}$, then as $n \to \infty$

$$\eta^2(f_n) = E[f_n(\vec{x}, a_n) - f(\vec{x})]^2 \sim \frac{a_n^p}{n} f(\vec{x}) \prod_{j=1}^{p} \int K_j^2(u)\,du + a_n^{-2s} \left(\frac{1}{s!} \sum_{j=1}^{p} \alpha_j \frac{\partial^s f(\vec{x})}{\partial x_j^s} \right)^2 \tag{2.4}$$

and the optimal value $a_n = a_n^0$ which minimizes the m.s.e. under the additional condition

$$B(\vec{x}) = \sum_{j=1}^{p} \alpha_j \frac{\partial^s f(\vec{x})}{\partial x_j^s} \neq 0$$

is equal to

$$a_n^0 = \left[\frac{2sB^2(\vec{x})}{(s!)^2 p f(\vec{x}) \tilde{K}_p} \right]^{1/(p+2s)} n^{1/(p+2s)} = Cn^{\gamma}, \tag{2.5}$$

where

$$\tilde{K}_p = \prod_{i=1}^{p} \int K_i^2(x)\,dx, \quad C^{2s+p} = C^{2s+p}(\vec{x}) = \frac{2sB^2(\vec{x})}{(s!)^2 p f(\vec{x}) \tilde{K}_p},$$

$$\gamma = \frac{1}{2s + p}.$$

For $a_n = a_n^0$

$$\eta^2(f_n) \sim (p + 2s)((2s)^{-1} f(\vec{x})\tilde{K}_p)^{2s/(2s+p)} \times$$
$$\times ((s!)^{-1} p^{-1/2} B(\vec{x}))^{2p/(p+2s)} n^{-2s/(p+2s)} \quad (2.6)$$

Thus the optimal estimator of the density $f_n(\vec{x}, a_n^0)$ is consistent in the mean square sense of order $N(n) = n^{-2s/(p+2s)}$ i.e. $N(n)\eta^2(f_n) \to$ to a finite limit.

3. We shall investigate the asymptotic behavior of the moments of the estimator of derivatives of the density $f(\vec{x})$, $x \in R_p$. It is natural to choose the statistic $\partial^m f_n(\vec{x})/\partial x_i^m$ as an estimator of $\partial^m f(\vec{x})/\partial x_i^m$.

Using arguments analogous to those presented in the course of the proof of Lemma 2.1 one proves:

LEMMA 2.2 (cf. also Theorem 4.5.1). If $f(\vec{x})$ satisfies conditions G_s, $K_j \in H_s$, $j = \overline{1, p}$ and moreover, if $K_j(x)$ possess bounded integrable derivatives of the s-th order and $\int x^s |K_j^{(s)}(x)|\, dx < \infty$ then

$$\lim_{n\to\infty} E \frac{\partial^s f_n(\vec{x}, a_n)}{\partial x_i^s} = \frac{\partial^s f(\vec{x})}{\partial x_i^s},$$

LEMMA 2.3. If $f(\vec{x})$ is bounded on R_p and $K_j(u)$, $j = \overline{1, p}$ possess bounded integrable derivatives of the m-th order, then as $n \to \infty$ and $r \geq 1$

$$E\left[\frac{\partial^m f_n(\vec{x}, a_n)}{\partial x_i^m} - E\frac{\partial^m f_n(\vec{x}, a_n)}{\partial x_i^m}\right]^{2r} = O\left(\left(\frac{a_n^{2m+p}}{n}\right)^r\right) \quad (2.7)$$

uniformly on the set of continuity points $C(f)$ of the function $f(\vec{x})$.

Proof. Introduce the notation

$$\psi(\vec{x}) = K_1(x_1) \cdot K_2(x_2) \ldots K_i^{(m)}(x_i) \ldots K_p(x_p),$$

$$\delta_n(\vec{x}, a_n) = n^{-1} a_n^p \sum_{j=1}^n \psi(a_n(\vec{x} - X_j)),$$

$$V_{n_i} = V(X_i) = a_n^p(\Psi(a_n(x - X_i)) - E\Psi(a_n(\vec{x} - X_i))) \times$$
$$\times (Da_n^p \psi(a_n(\vec{x} - X_j)))^{-1/2},$$
$$V_n = V(X).$$

We have

$$\xi_n = \left[\frac{\partial^m f_n(\vec{x}, a_n)}{\partial x_i^m} - E\frac{\partial^m f_n(\vec{x}, a_n)}{\partial x_i^m}\right]^{2r} =$$

$$= (a_n^m (D\delta_n(\vec{x}, a_n))^{1/2})^{2r} \left(\frac{\delta_n(\vec{x}, a_n) - E\delta_n(\vec{x}, a_n)}{\sqrt{D\delta_n(\vec{x}, a_n)}} \right)^{2r},$$

and moreover

$$(\delta_n(\vec{x}, a_n) - E\delta_n(\vec{x}, a_n))(D\delta_n(\vec{x}, a_n))^{-1/2} = n^{-1/2} \sum_{i=1}^{n} V_{n_i}.$$

In a manner analogous to Lemma 2.2 one can prove that as $n \to \infty$

$$\left(\frac{a_n^p}{n} \right)^{-1} D\delta_n(\vec{x}, a_n) \to f(x) \int \Psi^2(\vec{u})\, d\vec{u}. \tag{2.8}$$

On the other hand, in accordance with Theorem 1 of von Bahr [29] we have

$$E((\delta_n(\vec{x}, a_n) - E\delta_n(\vec{x}, a_n))(D\delta_n(\vec{x}, a_n))^{-1/2})^{2r} =$$

$$= \int u^{2r}\, d\Phi(u) + \sum_{j=1}^{2r-2} n^{-j/2} \int u^{2r}\, dP_{nj}(-\Phi)(u). \tag{2.9}$$

$P_{nj}(-\Phi)$, $j = 1, 2, \ldots$ are defined as in Section 1 of Chapter 1 and depend only on the moments $\alpha_n^{(i)} = E(V_n)^i$, $i = \overline{3, j+2}$.

Therefore

$$E\xi_n = D_{nr}(\vec{x}) \left[\int u^{2r}\, d\Phi(u) + \sum_{j=1}^{2r-2} n^{-j/2} \int u^{2r}\, dP_{nj}(-\Phi)(x) \right], \tag{2.10}$$

where

$$D_{nr}(\vec{x}) = \left(a_n^m \sqrt{D\delta_n(\vec{x}, a_n)} \right)^{2r}.$$

It is easy to verify that

$$D_{nr}(\vec{x}) = \left(\frac{a_n^m}{\sqrt{n}} \sqrt{Da_n^p \Psi(a_n(\vec{x} - X))} \right)^{2r} \leq C_1^r \left(\frac{a_n^{2m+p}}{n} \right)^r \times$$

$$\times \left[\int \Psi^2(\vec{u})\, d\vec{u} \right]^r = O\left(\left(\frac{a_n^{2m+p}}{n} \right)^r \right), \tag{2.11}$$

where $C_1 = \sup_x f(x)$.

Using analogous arguments one can verify that the following inequality is valid:

$$|D_{nr}(\vec{x})| \leq C_2 \left(\frac{a_n^{2m+p}}{n} \right)^r \left(\frac{a_n^p}{n} \right)^{(k-2)/2}, \tag{2.12}$$

where $k \geq 3$.

Relations (2.10)-(2.12) imply (2.7) which proves the lemma.

Next, taking (2.8), (2.11) and (2.12) into account, one can easily derive from (2.10) that

$$\left(\frac{n}{a_n^{2m+p}}\right)^r E\xi_n \to (f(x) \int \Psi^2(\vec{u})\, d\vec{u})^r \int u^{2r}\, d\Phi(u). \tag{2.13}$$

at all points of continuity of $f(\vec{x})$ (it is easy to verify that the odd order moments approach zero). In particular for $m = 0$ we have from (2.13) that

$$(na_n^{-p})^r E(f_n(\vec{x}, a_n) - Ef_n(\vec{x}, a_n))^{2r} \to$$

$$\to [f(\vec{x}) \prod_{j=1}^{p} \int K_j^2(u)\, du]^r \int u^{2r}\, d\Phi(u).$$

Thus the moments of the random variable $\zeta_n = (na_n^{-2m-p})^{1/2}\xi_n^{1/2r}$ as $n \to \infty$ converge to the moments of the normal distribution $\Phi_\sigma(\lambda)$, where $\sigma^2 = f(\vec{x}) \int \Psi^2(\vec{u})\, d\vec{u}$ ($\Phi_\sigma(\lambda)$ is the normal distribution function with mean zero and variance σ^2). Thus, under the conditions of Lemma 2.3 we have proved the following theorem.

THEOREM 2.2.

$$\zeta_n = (n^{-1}a_n^{2m+p})^{-1/2}\left(\frac{\partial^m f_n(\vec{x}, a_n)}{\partial x_i^m} - E\,\frac{\partial^m f_n(\vec{x}, a_n)}{\partial x_i^m}\right)$$

is distributed in the limit normally with mean 0 and variance $f(\vec{x}) \int \Psi^2(\vec{u})\, d\vec{u}$.

We shall now estimate the maximal derivation of $P(\sigma^{-1}\zeta_n < \lambda)$ from $\Phi(\lambda)$. We have

$$\Delta_n = \sup_{-\infty<\lambda<\infty} |P(\sigma^{-1}\zeta_n < \lambda) - \Phi(\lambda)| \leq \sup_{-\infty<\lambda<\infty} |P(\sigma_n^{-1}\zeta_n < \lambda) -$$

$$- \Phi(\lambda)| + \sup_{-\infty<\lambda<\infty} |\Phi_{\sigma_n^*}(\lambda) - \Phi(\lambda)|, \tag{2.14}$$

where

$$\sigma_n^2 = Ea_n^2\Psi^2(a_n(\vec{x} - X)) \sim \sigma^2 \quad \text{as} \quad n \to \infty \quad \text{and} \quad \sigma_n^* = \frac{\sigma_n}{\sigma}.$$

Utilizing the Berry-Esseen theorem we obtain that

$$\sup_{\lambda} |P(\sigma_n^{-1}\zeta_n < \lambda) - \Phi(\lambda)| \leq L_n^{(1)} = C_3 n^{-1/2} E|\tilde{V}_n|^3 (D\tilde{V}_n)^{-3/2}, \tag{2.15}$$

and moreover,

$$L_n^{(1)} \sim C_3(n^{-1}a_n^p)^{1/2} f^{-1}(\vec{x}) \int |\Psi(\vec{u})|^3\, d\vec{u} (\int \Psi^2(u)\, du)^{-3/2} \text{ as } n \to \infty,$$

where

$$\tilde{V}_n = a_n^p \Psi(a_n(\vec{x} - X)).$$

On the other hand, it is easy to verify that

$$\sup_{\lambda} |\Phi_{\sigma*}(\lambda) - \Phi(\lambda)| \leq L_n^{(2)} = 2\left(\left|\frac{\sigma_n^2}{\sigma^2} - 1\right|\right). \tag{2.16}$$

Thus, in view of (2.15) and (2.16) it follows from (2.14) that

$$\Delta_n \leq L_n^{(1)} + L_n^{(2)} = O((n^{-1}a_n^p)^{1/2}) + O(|\sigma^{-2}\sigma_n^2 - 1|).$$

4. We shall now proceed to a solution of the problem posed in the beginning of this section. Assume that $f \in G_s$ and $K_j \in H_s$ and, moreover, let them satisfy the conditions of the Lemma 2.3. Consider the optimal value $a_n = a_n^0$ obtained which yields the minimum of the m.s.e. Since the function $C = C(\vec{x})$ which appears in (2.5) is unknown we shall first estimate it from the sample X_i, $i = \overline{1, n}$. Let $\{\tau_{ni}\}$, $i = \overline{0, p}$ be sequences of positive numbers such that $\tau_{ni} \to \infty$, $i = \overline{1, p}$ as $n \to \infty$ and $\tau_{no} = O(n^{\nu})$, $\tau_{ni} = O(n^{\beta})$, $\beta = \gamma^2$, $j = \overline{1, p}$. Furthermore, let $\{b_n\}$ be a sequence of positive numbers converging to zero and satisfying the condition $nb_n^p \geq C_4 > 0$.

Introduce the notation:

$$f_n = f_n(\vec{x}, \tau_{no}), \qquad f_{nj}^{(s)} = \frac{\partial^s f_n(\vec{x}, \tau_{nj})}{\partial x_j^s},$$

$$\mu_n = E(f_n) \quad \text{and} \quad \mu_{nj}^{(s)} = E(f_{nj}^{(s)}).$$

The properties of the estimators f_n and $f_{nj}^{(s)}$ suggest that it would be appropriate to consider the sequence of estimators C^{2s+p} of the form

$$\tilde{C}_n^{2s+p} = 2s\left[\left(\sum_{i=1}^{p} \alpha_j f_{nj}^{(s)}\right)^2 + b_n\right]\left[(s!)^2 pK_p(|f_n| + b_n)\right]^{-1}. \tag{2.17}$$

It follows immediately from Lemmas 2.2 and 2.3 that $\tilde{C}_n$ is a consistent estimator of $C = C(\vec{x})$.

Set

$$C_n^{2s+p} = 2s\left[\left(\sum_{j=1}^{p} \alpha_s \mu_{nj}^{(s)}\right)^2 + b_n\right]\left[(s!)^2 p\tilde{K}_p(|\mu_n| + b_n\right]^{-1},$$

$$\lambda_n^2 = n^{-1} \max(\tau_{no}^p, \tau_{n1}^{2s+p}, \ldots, \tau_{np}^{2s+p}), \quad \tilde{\tau}_n = \tilde{C}_n n^{\gamma}, \quad \sigma_n = C_n n^{\gamma}.$$

We estimate $E|\tilde{C}_n - C_n|^m$, $m \geq 1$. Since $n^{-1}\tau_{no}^p = O(n^{-2s\gamma})$ and

$$n^{-1}\tau_{nj}^{2s+p} = O(n^{-(2s+q)}), \qquad q = p - 1 \geq 0,$$

it follows that $\lambda_n^2 = O(n^{-2s\gamma})$.

Taking this fact and relation (2.7) into account it is easy to show that

$$E|f_n - \mu_n|^m = O(\lambda_n^m), \quad E|f_{ni}^{(s)} - \mu_{ni}^{(s)}|^m = O(\lambda_n^m),$$

$$E\left|\left(\sum_{j=1}^{p} \alpha_j f_{nj}^{(s)}\right)^2 - \left(\sum_{j=1}^{p} \alpha_j \mu_{nj}^{(s)}\right)^2\right|^m = O(\lambda_n^m). \tag{2.18}$$

Furthermore, since $|f_n| + b_n \geq h_n = b_n/\tau_{no}^p$, $|\mu_n| + b_n \geq f(\vec{x}) - \varepsilon$ for $n > N(\varepsilon)$ and $E|f_n - \mu_n|^m = O(\lambda_n^m)$ and it follows from the inequality $2s(2p^2 - 1) > p(1 + p)$, $s \geq 2$, $p \geq 1$, that $\lambda_n^{4p} = O(h_n)$. We have in view of Lemma 1.1.2. that

$$E|(|f_n| + b_n)^{-1} - (|\mu_n| + b_n)^{-1}|^m = O(\lambda_n^m).$$

This, together with (2.18) implies that

$$E|\tilde{C}_n^{2s+p} - G_n^{2s+p}|^m = O(\lambda_n^m). \tag{2.19}$$

Consider once more Lemma 1.2. Setting in it $\eta_n = \tilde{C}_n^{2s+p}$

$$\zeta_n = C_n^{2s+p}, \quad d_n = sb_n(ps!\tilde{K}_p M\tau_{no}^p)^{-1}, \quad K = 4p,$$

$$\delta = 2^{-1}(C^{2s+p} - \varepsilon), \quad \alpha = (2s + p)^{-1},$$

in view of (2.19) we obtain

$$E|\tilde{C}_n - C_n|^m = O(\lambda_n^m), \tag{2.20}$$

where

$$M = \max\{\sup_x |K_j(x)|, \quad j = \overline{1, p}\}.$$

Finally we introduce the function

$$K^*(\vec{x}) = p\prod_{j=1}^{p} K_j(x_j) + \sum_{i=1}^{p}\left(\prod_{\substack{j=1 \\ j\neq 1}}^{p} K_j(x_j)\right) K_i^{(1)}(x_i)x_i$$

and prove the theorem which is a generalization of Theorem 1 [14].

THEOREM 2.3. Let $f(\vec{x}) \in G_s$, $K_j(u) \in H_s$, $j = \overline{1, p}$ and moreover, let $K_j(u)$ possess bounded, integrable derivatives of the s-th order and $\int x^s|K^{(s)}(x)|\,dx < \infty$. If $K^*(\vec{x})$ admits nonincreasing and integrable majorant $K_0(\vec{x})$ on $[0, \infty)^p$ then as $n \to \infty$

$$E(f_n(\vec{x}, \hat{\tau}_n) - f(x))^2 \sim E(f_n(\vec{x}, a_n^0) - f(\vec{x}))^2, \tag{2.21}$$

where

$$\hat{\tau}_n = \tilde{C}_n n^\gamma.$$

Proof. To prove (2.21) it is sufficient to verify that

$$E(f_n(\vec{x}, \hat{\tau}_n) - f_n(\vec{x}, \sigma_n))^2 = O(n^{-2s\gamma}). \qquad (2.22)$$

From the formula of finite increments we obtain

$$f_n(\vec{x}, \hat{\tau}_n) - f_n(\vec{x}, \sigma_n) = \left(\frac{\partial f_n(\vec{x}, t)}{\partial t}\right)_{t=\xi_n} (\hat{\tau}_n - \sigma_n),$$

where the random variable ξ_n is located between $\hat{\tau}_n$ and σ_n and

$$E(f_n(\vec{x}, \hat{\tau}_n) - f_n(\vec{x}, \sigma_n))^2 \leq$$

$$\leq \left[E\left(n^\gamma\left(\frac{\partial f_n(x, t)}{\partial t}\right)_{t=\xi_n}\right)^4 E(\hat{C}_n - C_n)^4\right]^{1/2}.$$

From formula (2.20) we have in particular that

$$(E(\tilde{C}_n - C_n)^4)^{1/2} = O(\lambda_n^2) = O(n^{-2s\gamma}).$$

Therefore it remains to verify that

$$E\left(n^\gamma\left(\frac{\partial f_n(\vec{x}, t)}{\partial t}\right)_{t=\xi_n}\right)^4 = O(1).$$

Let A_n be the event that $\tilde{C}_n \geq C - \varepsilon$, $\varepsilon > 0$, and I_{A_n} be the indicator of the set A_n. Furthermore, let $\eta_n = (C - \varepsilon)n^\gamma$. We have

$$E\left(n^\gamma\left(\frac{\partial f_n(\vec{x}, t)}{\partial t}\right)_{t=\xi_n}\right)^4 \leq E\left[n^\gamma \frac{\xi_n^{p-1}}{n} \sum_{i=1}^n |K^*(\xi_n(\vec{x} - X_i))\, I_{A_n}\right]^4 +$$

$$+ E\left[n^\gamma \frac{\xi_n^{p-1}}{n} \sum_{i=1}^n |K^*(\xi_n(\vec{x} - X_i))|\, I_{\bar{A}_n}\right]^4 \leq n^{4\gamma}(I_n^{(1)} + I_n^{(2)}).$$

We bound $n^{4\gamma}I_n^{(1)}$. Since $\xi_n \geq (C - \varepsilon)n^\gamma$ on the set A_n, we have

$$n^{4\gamma}I_n^{(1)} \leq \frac{1}{(C - \varepsilon)^4} E\left\{\left[\frac{\xi_n^p}{\eta_n^p} \frac{\eta_n^p}{n} \sum_{i=1}^n K_0(\eta_n(\vec{x} - X_i))\right] I_{A_n}\right\}^4 \leq$$

$$\leq C_5 \left\{E\left(\frac{\xi_n}{\eta_n}\right)^{8p} E\left[\frac{\eta_n^p}{n} \sum_{i=1}^n K_0(\eta_n(\vec{x} - X_i))\right]^8\right\}^{1/2}. \qquad (2.23)$$

Now, in view of Lemma 2.3 one can write

$$E\left[\frac{\eta_n^p}{n} \sum_{i=1}^n K_0(\eta_n(\vec{x} - X_i))\right]^8 = O(1). \qquad (2.24)$$

On the other hand,

$$E\left(\frac{\xi_n}{\eta_n}\right)^8 \leq C_6 + 2^{8p}E\left|\frac{\sigma_n - \hat{\tau}_n}{\eta_n}\right|^{8p} = C_6 + C_7E\left|C_n - \tilde{C}_n\right|^{8p} =$$

$$= C_6 + O(n^{-2s}). \tag{2.25}$$

In view of relations (2.23)-(2.25) we conclude that $n^{4\gamma}I_n^{(1)}$ is bounded.

Finally

$$I_n^{(2)} \leq M^4E[\xi_n^{p-1}I_{\bar{A}_n}]^4 \leq C_9P(\bar{A}_n), \tag{2.26}$$

where $M = \max_x K_0(x)$.

For n sufficiently large, $C_n > C - \varepsilon/2$, therefore utilizing Chebyshev's inequality we have from (2.20) that

$$P(\bar{A}_n) = O(n^{-2sp\gamma}). \tag{2.27}$$

Substituting (2.27) into (2.26) we obtain that $n^{4\gamma}I_n^{(2)}$ is also bounded. The theorem is thus proved.

Chapter 2

STRONGLY CONSISTENT IN FUNCTIONAL METRICS ESTIMATORS OF PROBABILITY DENSITY

1. Strong Consistency of Kernel-Type Density Estimators in the Norm of the Space C

1. Let X_j, $j = 1, 2, \ldots$ be a sequence of independent, identically distributed random variables (observations) with values in R_1. Let their common distribution function $F(x)$ be absolutely continuous and $f(x)$ be the corresponding density function. As an estimator of the density $f(x)$ we consider a non-parametric estimator $f_n(x)$ of the form (cf. 1.1.1)

$$f_n(x) = a_n n^{-1} \sum_{i=1}^{n} K(a_n(x - X_i)),$$

where $K(x)$ is a probability density and $\{a_n\}$ is a sequence of positive numbers approaching infinity ($a_n \uparrow \infty$ as $n \to \infty$).

In this section we shall investigate the problem of deviations of density estimators $f_n(x)$ from the density $f(x)$ in the uniform metric. In his groundbreaking paper, E. Parzen [5] has shown that $\rho(f_n, f) = \sup_x |f_n(x) - f(x)|$ converges to zero in probability provided the Fourier transform of the kernel $K(x)$ is absolutely integrable, $na_n^{-2} \to \infty$ and the density $f(x)$ is uniformly continuous. This result is mainly of a qualitative nature but admits refinements in various directions. One such refinement (cf. also Section 1 in Chapter 3) consists of a solution of the following problem (due to N. V. Smirnov): under what conditions on $f(x)$, $K(x)$ and a_n, does the estimator $f_n(x)$ converge almost surely (a.s.) to $f(x)$ uniformly on the whole real line?

The conditions for a.s. convergence are given in

THEOREM 1.1. If
1. $K(x)$ is a function of bounded variation;
2. The series $\sum_{n=1}^{\infty} e^{-\gamma n a_n^{-2}}$ converges for any $\gamma > 0$;
3. $f(x)$ is uniformly continuous on the whole real axis, then $\rho(f_n, f) = \|f_n - f\|_C$ converges to zero a.s. as $n \to \infty$.

<u>Proof</u>. We shall show that $\sup_{x \in R_1} |f_n(x) - Ef_n(x)| \to 0$ a.s.

Indeed, applying the formula of integration by parts we obtain that

$$\sup_{x \in R_1} |f_n(x) - Ef_n(x)| = \sup_{x \in R_1} \Big| a_n \int K(a_n(x - u))\, dF_n(u) -$$

$$- a_n \int K(a_n(x - u))\, dF(u)| \leq \sup_{x \in R_1} \int a_n| F_n(u) -$$

$$- F(u)|\, dK(a_n(x - u))| \leq \sup_{u \in R_1} |F_n(u) - F(u)|\, a_n \cdot \mu, \qquad (1.1)$$

where $\mu = \text{Var}(K)$, $F_n(x)$ is the empirical distribution function constructed from the sample.

As usual, let

$$D_n = \sup_{x \in R_1} |F_n(x) - F(x)|, \qquad D_n^+ = \sup_{x \in R_1} (F_n(x) - F(x)),$$

$$D_n^- = \sup_{x \in R_1} (F(x) - F_n(x)).$$

It is easy to verify that

$$P\left\{D_n > \frac{\lambda}{\sqrt{n}}\right\} \leq P\left\{ D_n^+ > \frac{\lambda}{\sqrt{n}} \right\} + P\left\{ D_n^- > \frac{\lambda}{\sqrt{n}} \right\} = 2(1 - P_n^+(\lambda)), \qquad (1.2)$$

where $P_n^+(\lambda) = P\{D_n^+ < \lambda n^{-1/2}\}$. From the well-known Smirnov's formula [35] follows the inequality

$$1 - P_n^+(\lambda) \leq C_1 e^{-2\lambda^2}, \qquad (1.3)$$

here and below C_i, $i \geq 1$ denotes a positive constant. Then from (1.2) and (1.3) we have

$$P\{D_n > \lambda n^{-1/2}\} \leq C_2 e^{-2\lambda^2}. \qquad (1.4)$$

In view of (1.1) and (1.4)

$$P\{ \sup_{x \in R_1} |f_n(x) - Ef_n(x)| > \varepsilon\} \leq P \sup_{x \in R_1} |F_n(x) - F(x)| >$$

$$> \varepsilon a_n^{-1} \mu^{-1}\} \leq C_2 e^{-\beta n a_n^{-2}}, \qquad (1.5)$$

where

$$\beta = 2 \left(\frac{\varepsilon}{\mu}\right)^2 .$$

Applying the Borel-Cantelli lemma to the inequality (1.5) we can assert that $\sup_{x \in R_1} |f_n(x) - Ef_n(x)|$ approaches zero a.s. as $n \to \infty$.

To prove the theorem it remains to show that $\sup_{x \in R_1} |Ef_n(x) - f(x)| \to 0$. Let $\delta > 0$. Then setting $M = \max_{x \in R_1} f(x)$, we have

$$\sup_{x \in R_1} |Ef_n(x) - f(x)| \leq \sup_{x \in R_1} \int_{|y| \leq \delta} | f(x - y) - f(x) |\, a_n K(a_n y)\, dy +$$

$$+ \sup_{x \in R_1} \int_{|y|>\delta} |f(x - y) - f(x)| a_n K(a_n y)\, dy \leq$$

$$\leq \sup_{x \in R_1} \sup_{|y|\leq\delta} |f(x - y) - f(x)| + 2M \int_{|y|>\delta a_n} K(y)\, dy. \quad (1.6)$$

Let $\eta > 0$ be arbitrarily small. By choosing δ sufficiently small the first summand on the r.h.s. of (1.6) can be made smaller than $\eta/2$ since $f(x)$ is uniformly continuous; for a fixed δ one can select n so large that the second term on the r.h.s. of (1.6) will also be smaller than $\eta/2$. Then it follows from (1.6) that

$$\sup_{x \in R_1} |Ef_n(x) - f(x)| < \eta.$$

The theorem is thus proved.

Now let $K(x)$ be continuous and $K(x) \to 0$ as $|x| \to \infty$. Then there exists a random variable Θ_n such that

$$f_n(\Theta_n) = \max_x f_n(x).$$

We call Θ_n the <u>sample mode</u> of the distribution $f(x)$.

THEOREM 1.2. Let the conditions 1°-3° of Theorem 1.1 be fulfilled and moreover, let $K(x)$ be continuous and $K(x) \to 0$ as $|x| \to \infty$. Then if the theoretical mode Θ defined by the equality $f(\Theta) = \max_{x \in R_1} f(x)$ is unique, the sample mode Θ_n converges to Θ a.s.:

$$\Theta_n \to \Theta \text{ a.s.}$$

<u>Proof</u>. Since $f(x)$ is a uniformly continuous density with a unique mode Θ it has the following properties:

1) for any $\varepsilon > 0$ there exists a $\eta > 0$ such that $|\Theta - x| \geq \varepsilon$ implies $|f(\Theta) - f(x)| \geq \eta$;

2) $|f(\Theta_n) - f(\Theta)| \leq 2\rho(f_n, f)$. These properties and Theorem 1.1 yield the proof of the theorem.

<u>Remark 1</u>. The series $\sum_{n=1}^{\infty} \exp(-\gamma n a_n^{-2})$, $\gamma > 0$, is a Dirichlet series of the type $\lambda_n = n a_n^{-2}$. The series $\sum_{n=1}^{\infty} \exp(-\gamma n a_n^{-2})$ converges if and only if $(a_n^2 \log n)/n \to 0$ as $n \to \infty$.

2. It should be noted that the conditions of Theorem 1.1 are in a certain sense the best. Schuster has proved [17] the following theorem: Let $K(u)$ satisfy condition 1° and some additional conditions and a_n satisfy condition 2°. In order that $\lim_{n\to\infty} \rho(f_n, g) = 0$ a.s. it is necessary and sufficient that the function $g(x)$ be a uniformly continuous density of the distribution $F(x)$ with respect to the Lebesgue measure.

It turns out that the additional conditions on $K(u)$ are not necessary. We shall present a brief proof of this theorem under conditions 1° and 2°.

Sufficiency follows from Theorem 1.1. We shall now prove its necessity. In view of (1.5) $\rho(f_n, Ef_n) \to 0$ a.s. and the condition $\rho(f_n, g) \to 0$ a.s. Then as $n \to \infty$

$$\rho(Ef_n, g) \to 0. \tag{1.7}$$

We shall now show that F(x) is continuous. Assume the contrary, i.e. there exists a point x_0 such that $p_0 = F(x_0^+) - F(x_0) > 0$. Let $K(x_0') > 0$ and $x_n = a_n^{-1}x_0' + x_0$. Then

$$\sup_{x \in R_1} Ef_n(x) = \sup_{x \in R_1} \int a_n K(a_n(x - u))\, dF(u) \geq a_n \int K(a_n(x_n - u)) \times$$

$$\times\, dF(u) \geq a_n K(a_n(x_n - x_0))p_0 = p_0 a_n K(x_o') \tag{1.8}$$

On the other hand

$$\sup_{x \in R_1} Ef_n(x) \leq a_n \cdot \mu, \quad \mu = \mathrm{Var}(K). \tag{1.9}$$

Inequalities (1.8) and (1.9) contradict relation (1.7). Thus F(x) is continuous.

It follows from this assertion that $Ef_n(x)$ is uniformly continuous. Hence in view of (1.7) g(x) is also uniformly continuous and

$$\int_a^x Ef_n(x)\, dx \to \int_a^x g(x)\, dx \quad \text{as} \quad n \to \infty.$$

On the other hand, it is easy to verify that

$$\int_a^x Ef_n(x)\, dx \to F(x) - F(a) \quad \text{as} \quad n \to \infty.$$

Hence,

$$F(x) - F(a) = \int_a^x g(u)\, du.$$

The theorem is thus proved.

Remark 2. Let a sample X_j, $j = \overline{1, n}$ be given from a p-dimensional population distributed according to unknown probability density $f(\vec{x})$, $\vec{x} = (x_1, \ldots, x_p)$. To estimate $f(\vec{x})$ consider the statistic (1.2.1) with an equal step: $a_i = a_n$, $i = \overline{1, p}$. All the results obtained in this section can be extended to the multivariate case.

The following theorem is valid.

THEOREM 1.3. If

1°. $K_j(u)$, $j = \overline{1, p}$, is a function of bounded variation;

2°. The series $\sum_{n=1}^{\infty} e^{-\gamma\lambda_n}$, $\lambda_n = na_n^{-2p}$ is convergent for any $\gamma > 0$;

3°. $f(x)$ is uniformly continuous on R_p, then

$$\sup_{x \in R_p} |f(\vec{x}) - f_n(\vec{x})| \to 0 \text{ a.s.}$$

The proof of this theorem is analogous to the corresponding proof in one dimensional case. For this purpose, only an analog of the inequality (1.4) is required which is valid in the multidimensional case as well [36]:

$$P\{\sqrt{n} \sup_{\vec{x} \in R_p} |F_n(\vec{x}) - F(\vec{x})| > \lambda\} \leq C_3 e^{-C_4\lambda^2}, \tag{1.10}$$

where C_3 and C_4 are finite constants which depend on the dimensionality of the Euclidean space with $0 < C_4 \leq 2$ and $F_n(\vec{x})$ is the p-dimensional empirical distribution function.

3. In this subsection the problem of estimating the parameter $\Theta(f) = \sup_{x \in R_1} f(x)$ is considered.

A natural estimator for the quantity $\Theta(f)$ is

$$\hat{\Theta}_n = \sup_{x \in R_1} f_n(x). \tag{1.11}$$

Denote by $W(\beta, L)$ ($\beta = m + \alpha$, $0 < \alpha \leq 1$, $m \geq 0$) the set of bounded and m times differentiable functions such that for $x_1, x_2 \in R_1$

$$|f^{(m)}(x_2) - f^{(m)}(x_1)| \leq L|x_2 - x_1|^{\alpha}. \tag{1.12}$$

THEOREM 1.4. For the estimator of the form (1.11) with $a_n = n^{1/[2(\beta+1)]}$, $\beta = m + \alpha$ and the function $K(x)$ satisfying the conditions:

1° $\int K(x)\,dx = 1$, $\int x^i K(x)\,dx = 0$, $i = \overline{1, m}$,

$x^{\beta} K(x) \in L_1(-\infty, \infty)$,

2° $K^*(t) = \int e^{itx} K(x)\,dx \in L_1(-\infty, \infty)$,

the relation

$$\sup_{f \in W(\beta,L)} E(\hat{\Theta}_n - \Theta(f))^2 \leq C_5 n^{-\beta/(1+\beta)},$$

is valid where C_5 depends only on L, $A(\beta) = \int |x^{\beta} K(x)|\,dx$, and $\int |K^*(t)|\,dt$.

Proof. Since

$$|\hat{\Theta}_n - \Theta(f)| = |\sup_{x \in R_1} f_n(x) - \sup_{x \in R_1} f(x)| \leq \sup_{x \in R_1} |f_n(x) - f(x)|,$$

it follows that

$$E(\hat{\Theta}_n - \Theta(f))^2 \leq 2E(\sup_{x \in R_1} |f_n(x) - Ef_n(x)|)^2 +$$

$$+ \left(\sup_{x \in R_1} |Ef_n(x) - f(x)|\right)^2). \tag{1.13}$$

Using the Taylor expansion of the function $f(x)$ and the condition 1° of the theorem we obtain

$$Ef_n(x) - f(x) = \frac{a_n^{-m}}{m!} \int y^m K(y)[f^{(m)}(\xi) - f^{(m)}(x)]\, dy,$$

where ξ is a point belonging to the interval $(x - a_n^{-1}y, x)$. Now, applying (1.12) we verify the inequality

$$\sup_{x \in R_1} |Ef_n(x) - f(x)| \le C_6 a_n^{-\beta}, \tag{1.14}$$

where C_6 depends on L and $A(\beta)$.

Next, we have from condition 2° that

$$f_n(x) = \frac{1}{2\pi} \int e^{-iux} K^* \left(\frac{u}{a_n}\right) \phi_n(u)\, du,$$

where

$$\phi_n(u) = \frac{1}{n} \sum_{j=1}^{n} e^{iux_j}.$$

Hence

$$\left(\sup_{x \in R_1} |f_n(x) - Ef_n(x)|\right)^2 \le \frac{1}{4\pi^2} \int [\phi_n(t) - \phi(t)]^2 \left| K^* \times \right.$$

$$\left. \times \left(\frac{t}{a_n}\right)\right| dt \int \left|K^* \left(\frac{t}{a_n}\right)\right| dt,$$

where $\phi(t)$ is the characteristic function of the random variable X_1. Hence

$$E\left(\sup_{x \in R_1} |f_n(x) - Ef_n(x)|\right)^2 \le \frac{a_n^2}{n} \frac{1}{4\pi^2} \left(\int |K^*(t)|\, dt\right)^2. \tag{1.15}$$

Choosing $a_n = n^{1/[2(1+\beta)]}$ we arrive from (1.13), (1.14) and (1.15) at the assertion of the theorem.

2. Convergence in the L_2 Norm of Kernel-Type Density Estimators

1. Let X_i, $i \ge 1$, be independent, identically distributed p-dimensional, $p \ge 1$, random vectors with the probability density $f(\vec{x})$, $\vec{x} \in R_p$. We shall consider sequences of estimators $f_n(\vec{x})$, $n \ge 1$,

$$f_n(\vec{x}) = n^{-1} a_n^p \sum_{i=1}^{n} K(a_n(\vec{x} - X_i)), \tag{2.1}$$

which generalize the estimators (1.2.1) where $K(\vec{x})$ is a probability density satisfying the conditions:

$$\sup_{\vec{u} \in R_p} K(\vec{u}) < \infty,$$

$$|\vec{u}|^p K(\vec{u}) \to 0 \quad \text{for} \quad |\vec{u}|^2 = u_1^2 + \ldots + u_p^2 \to \infty,$$

and $\{a_n\}$ is a sequence of positive numbers the same as the one defined in Section 1 of Chapter 2.

In the preceding section we measured the deviation of the estimator $f_n(\vec{x})$ from the density $f(\vec{x})$ in a uniform metric. Since a density is defined only up to the values taken on a set of measure zero it is more natural instead of the uniform distance to consider the distance in the metric of the L_2 space. Thus, we shall use the norm in the space L_2 of deviation of $f_n(\vec{x})$ from the true $f(\vec{x})$:

$$\rho_2(f_n, f) = \| f_n - f \|_2.$$

as the measure of the error of the estimator $f_n(\vec{x})$.

LEMMA 2.1. Let $K(x)$ satisfy the above stated conditions $f(\vec{x}) \in L_2(R_p)$ and $n^{-1} a_n^p \to 0$ as $n \to \infty$. Then

1° $$\| Ef_n \|_2^2 \le \| f \|_2^2,$$

2° $$\lim_{n\to\infty} \int Ef_n(\vec{x}) f(\vec{x})\, d\vec{x} = \int f^2(\vec{x})\, d\vec{x}.$$

Proof. It is easy to verify that

$$[Ef_n(\vec{x})]^2 \le a_n^p \int K(a_n(\vec{x} - \vec{u})) f^2(\vec{u})\, d\vec{u}, \tag{2.2}$$

$$E \int f_n(\vec{x}) f(\vec{x})\, d\vec{x} \le \left(\int f^2(\vec{x})\, d\vec{x}\right)^{1/2} \left(\int f^2(\vec{x} - a_n^{-1}\vec{z}) K(\vec{z})\, d\vec{z}\, d\vec{x}\right)^{1/2}. \tag{2.3}$$

Utilizing Fubini's theorem we obtain 1° from (2.2) and from (2.3) the inequality $E \int f_n(\vec{x}) f(\vec{x})\, d\vec{x} \le \int f^2(\vec{x})\, d\vec{x}$.

Consequently,

$$\overline{\lim_{n\to\infty}} \int Ef_n(\vec{x}) f(\vec{x})\, d\vec{x} \le \int f^2(\vec{x})\, d\vec{x}. \tag{2.4}$$

On the other hand, $\lim_{n\to\infty} Ef_n(\vec{x}) = f(\vec{x})$ at the point of continuity of $f(\vec{x})$ (cf. [33]). Therefore in view of Fatou's lemma

$$\int f^2(\vec{x})\, d\vec{x} \le \underline{\lim}_{n\to\infty} \int Ef_n(\vec{x}) f(\vec{x})\, d\vec{x}. \tag{2.5}$$

Assertion 2° now follows from (2.4) and (2.5).

LEMMA 2.2. If $f(\vec{x})$ and $K(\vec{x})$ satisfy the conditions of Lemma 2.1 and the series $\sum_{n=1}^{\infty} n \exp(-\gamma n a_n^{-2p})$ is convergent for any $\gamma > 0$ then, as $n \to \infty$, $\| f_n - Ef_n \|_2^2 \to 0$ a.s.

Proof. Introduce the notation

$$a_n^p K(a_n \vec{u}) * a_n^p K(a_n \vec{u}) = H_n'(\vec{u})$$

$$a_n^p K(a_n \vec{u}) * Ef_n(\vec{u}) = H_n''(\vec{u}),$$

where * denotes the convolution. It is easy to verify that

$$\| f_n - Ef_n \|_2^2 = n^{-1} \sum_{i=1}^{n} [n^{-1} \sum_{j=1}^{n} H_n'(X_j - X_i) - H_n''(X_i)] -$$

$$- n^{-1} \sum_{i=1}^{n} [H_n''(X_i) - EH''(X_i)],$$

moreover in view of the fact that $\sup_{\vec{u}} K(\vec{u}) < \infty$,

$$\| f_n - Ef_n \|_2^2 \leq C_5(na_n^{-p})^{-1} + \Sigma_n' + \Sigma_n'', \tag{2.6}$$

where

$$\Sigma_n' = n^{-1} \sum_{i=1}^{n} \left(\left| n^{-1} \sum_{\substack{j=1 \\ j \neq i}} H_n'(X_j - X_i) - H_n''(X_i) \right| \right),$$

and

$$\Sigma_n'' = n^{-1} \left| \sum_{i=1}^{n} (H_n''(X_i) - EH_n''(X_i)) \right|.$$

We shall now deal with the upper and lower bounds on the probabilities $P(\Sigma_n' \geq \varepsilon)$ and $P(\Sigma_n'' \geq \varepsilon)$ where $\varepsilon > 0$. For this purpose the following theorem due to Yu. V. Prohorov [37] is used.

Let $X_1, \ldots, X_n$ be independent, identically distributed m-dimensional random vectors, $EX_j = 0$, $|X_j| \leq L$, $E|X_i|^2 = \Delta$; then for $n \geq m$ the inequality

$$\vec{P}(|Y_n| \geq r) \leq C_0 \exp\left(-\frac{r^2}{8e^2L^2}\right), \tag{2.7}$$

is valid; here $Y_n = n^{-1/2}(X_1 + \ldots + X_n)$, $|\cdot|$ is the Euclidean norm of an m-dimensional vector and $C_0 = 1 + e^{5/12}/\pi\sqrt{2} \cdot \Delta/L^2$.

We bound $P(\Sigma_n'' \geq \varepsilon)$. Firstly, in view of the assumption $\sup_{x \in R_p} K(\vec{x}) \leq \leq C_6 < \infty$ we have

$$E(H_n''(X_i) - EH_n''(X_i))^2 \leq 4C_6^2 a_n^{2p},$$

$$|H_n''(X_i) - EH_n''(X_i)| \leq 2C_6 a_n^p.$$

Therefore, noting that in our case m = 1 and $C_0 \leq 2$, we obtain from (2.7) that

$$P(\Sigma_n'' \geq \varepsilon) \leq 2\exp(-\gamma_1 n a_n^{-2p}), \tag{2.8}$$

where $\gamma_1 = \varepsilon^2/32 \cdot e^2 C_6^2$.

We now proceed to estimate $P(\Sigma_n' \geq \varepsilon)$. Introduce the notation

$$\Sigma_n''' = n^{-1} \sum_{j=2}^{n} [H_n'(X_j - X_1) - H_n''(X_1)].$$

It is easily verified that

$$E([H_n'(X_j - X_1) - H_n''(X_1)]/X_1) = 0,$$

$$|H_n'(X_j - X_1) - H_n''(X_1)| \leq 2C_6 a_n^p, \quad j = \overline{2, n} \quad \text{and}$$

$$P(\Sigma_n' \geq \varepsilon) \leq nP(|\Sigma_n'''| \geq \varepsilon) = nEP(|\Sigma_n'''| \geq \varepsilon/X_1). \tag{2.9}$$

For a given X_1, $H_n'(X_j - X_1) - H_n''(X_1)$, $j = \overline{2, n}$, are independent, identically distributed random variables. Therefore utilizing Prohorov's inequality (2.7) we arrive at

$$P(|\Sigma_u'''| \geq \varepsilon/X_1) \leq 2\exp\left(-\frac{\varepsilon^2 n a_n^{-2p}}{8e^2 4C_6^2}\right) = 2\exp(-\gamma n a_n^{-2p}),$$

from which in view of (2.9) one can conclude that

$$P(\Sigma_n' \geq \varepsilon) \leq 2n\exp(-\gamma n a_n^{-2p}). \tag{2.10}$$

Since by the condition the series $\sum_{n=1}^{\infty} n\exp(-\gamma n a_n^{-2p})$ converges for any $\gamma > 0$ and it follows from (2.8) and (2.10) that Σ_n' and Σ_n'' tend to zero a.s. The lemma is thus proved.

The following theorem follows directly from Lemmas 2.1 and 2.2.

THEOREM 2.1. Let

$$\sup_{u \in R_p} K(\vec{u}) < \infty, \quad |\vec{u}|^p K(\vec{u}) \to 0 \quad \text{for} \quad |\vec{u}| \to \infty \quad \text{and}$$

$$f(\vec{x}) \in L_2(R_p).$$

Then if the series $\sum_{n=1}^{\infty} n\exp(-\gamma n a_n^{-2p})$ converges for any $\gamma > 0$ it follows that

$$\rho_2(f_n, f) \to 0 \quad \text{a.s.}$$

Remark. The Dirichlet series $\sum_{n=1}^{\infty} n\exp(-\gamma\lambda_n)$ of the type $\lambda_n = n a_n^{-2p}$ converges if and only if $(\log n)/\lambda_n \to 0$.

2. In some nonparametric derivations the functional $\alpha(f) = \int f^2(\vec{x})\, d\vec{x}$ is of importance. For example, this functional serves as the basic quantity in expressions for asymptotic efficiency of rank tests of Wilcoxon and van der Waerden types for testing hypotheses of location, regression, dependence, etc. Moreover, point estimators obtained from rank tests have asymptotic efficiency which include the functional $\alpha(f)$ which is general unknown. It is natural to use $\alpha(f_n)$ as an estimator of $\alpha(f)$. It is easy to evaluate the functional $\alpha(f_n)$. We present a simple example. Let $p = 1$ and $K(x) = \frac{1}{2}(0)$ respectively for $|x| \leq 1$ ($|x| > 1$).

We obtain

$$\alpha(f_n) = (2na_n^{-1})^2\left[2na_n^{-1} + 2 \sum_{i,j}{}' (2a_n^{-1} - |X_j - X_i|)\right],$$

where Σ' is the sum over all $1 \leq i < j \leq n$ such that $|X_j - X_i| \leq 2a_n^{-1}$.

The following strong assertion can be deduced from Theorem 2.1 just proved.

THEOREM 2.2. Under the conditions of the theorem $\alpha(f_n) \to \alpha(f)$ a.s. as $n \to \infty$.

We shall now investigate the asymptotic distribution of the statistic $\alpha(f_n)$ for $p = 1$ and will apply the result obtained herein to construction of confidence intervals and for testing hypotheses about the scale parameter of a distribution.

We have

$$\alpha(f_n) = \left(1 - \frac{1}{n}\right) U_n + \frac{a_n}{n} W(0), \tag{2.11}$$

where

$$U_n = \frac{1}{n(n-1)} \sum_{i \neq j}^{n} a_n W(a_n(X_i - X_j)) \quad \text{and } W = K*K.$$

Let $f(x)$ be bounded and possess bounded derivatives up to the second order inclusively. Moreover, let $x^2K(x) \in L_1(-\infty, \infty)$,

$$\frac{a_n}{n} \to 0 \quad \text{and } na_n^{-4} \to 0 \quad \text{as} \quad n \to \infty.$$

It is easy to verify that

$$\Theta_n = EU_n = a_n \iint W(a_n(u - v))f(u)f(v)\, du\, dv = \alpha(f) + O(a_n^{-2}). \tag{2.12}$$

Following W. Hoeffding [38] we introduce the notation

$$\phi_n(x_1) = E\{a_n W(a_n(x_1 - X_2))\}, \quad \Psi_n(x_1) = \phi_n(x_1) - \Theta_n,$$

$$\zeta_n^{(1)} = E\{\phi_n(X_1)\}^2 - \Theta_n^2, \quad \zeta_n^{(2)} = a_n^2 EW^2(a_n(X_1 - X_2)) - \Theta_n^2.$$

In view of formula (5.13) in [38] we can write

$$\sigma_n^2(f) = DU_n = \frac{4(n-2)}{n(n-1)}\zeta_n^{(1)} + \frac{2}{n(n-1)}\zeta_n^{(2)},$$

with $\zeta_n^{(2)} = O(a_n)$.

Next, utilizing the Lebesgue theorem on limiting transition under the sign of an integral we have as $n \to \infty$

$$\zeta_n^{(1)} = \int [\int a_n W(a_n(u - v))f(v)\, dv]^2 f(u)\, du - \Theta_n^2 \to$$

$$\to \int f^3(x)\, dx - \alpha^2(f) = Df(X_1).$$

Hence

$$n\sigma_n^2(f) \to \sigma^2(f) = 4[\int f^3(x)\, dx - \alpha^2(f)] \quad \text{as } n \to \infty. \tag{2.13}$$

Consider now the normalized sums of independent and indentically distributed random variables $\Psi_n(X_i)$, $i = 1, 2, \ldots;$

$$Y_n = \frac{\sum_{i=1}^{n} \Psi_n(X_i)}{\sqrt{nD\Psi_n(X_1)}}$$

and set

$$Z_n = \frac{U_n - \Theta_n}{\sqrt{DU_n}}.$$

Following W. Hoeffding [38] we have

$$E(Z_n - Y_n)^2 = 2 - 2E(Y_n \cdot Z_n),$$

and moreover

$$E(Y_n \cdot Z_n) = \frac{1}{2\sqrt{D\Psi_n(X_1)}} \frac{1}{\sqrt{nDU_n}} E \frac{2\sum_{i=1}^{n} \Psi_n(X_i)}{\sqrt{n}} \sqrt{n}(U_n - \Theta_n) =$$

$$= \frac{1}{\sqrt{D\Psi_n(X_1)}} \frac{2}{\sqrt{nDU_n}} \zeta_n^{(1)} \to 1 \quad \text{as} \quad n \to \infty.$$

Thus $E(Z_n - Y_n)^2 \to 0$ as $n \to \infty$.

We shall now verify that Y_n is distributed asymptotically normal (0.1). To do this it is sufficient to show that as $n \to \infty$

$$nP\left\{ \left| \frac{\Psi_n(X_i)}{\sqrt{D\Psi_n(X_1)}} \right| \geq \varepsilon n^{1/2} \right\} \to 0, \tag{2.14}$$

and to establish (2.14) it is sufficient to show that

$$\frac{E|\Psi_n(X_1)|^{2+\delta}}{n^{\delta/2} D\Psi_n(X_1))^{1+\delta}} \to 0, \quad \text{as} \quad n \to \infty \tag{2.15}$$

where $\delta > 0$.

(2.15) is fulfilled since as $n \to \infty$

$$D\Psi_n(X_1) \to \int f^3(x)\, dx - \alpha^2(f)$$

and

$$E|\psi_n(X_1)|^{2+\delta} \le C_7[\int |\phi_n(t)|^{2+\delta} f(t)\, dt + \Theta_n^{2+\delta}] \sim$$

$$\sim C_7[\int f^{2+\delta}(x)\, dx + \alpha^{2+\delta}(f)].$$

Thus the random variable Y_n is asymptotically distributed normal (0,1). Hence $Z_n = (U_n - \Theta_n)(DU_n)^{-1/2}$ is also distributed in the limit normal (0,1). From here taking (2.11), (2.12) and (2.13) into account we can state the following theorem.

THEOREM 2.3. Let $f(x)$ and $K(x)$ satisfy the above stated conditions. Then if $a_n/n \to 0$ and $na_n^{-4} \to 0$ as $n \to \infty$ the quantity

$$\sqrt{n}\, \frac{\alpha(f_n) - \alpha(f)}{\sigma(f)}$$

is distributed asymptotically normal (0,1).

Finally we shall consider certain statistical applications of this theorem.

Let the probability density $f(x)$ be a scale density of the form

$$f(x) = \gamma g(\gamma x), \quad 0 < \gamma < \infty,$$

where $g(x)$ is some known bounded probability density which possesses bounded derivatives up to the second order inclusively.

The statistic

$$\hat{\gamma}_n = \frac{\alpha(f_n)}{\alpha(g)}.$$

can be chosen as an estimator of the parameter γ.

It now follows from Theorem 2.2 that $\hat{\gamma}_n$ is a consistent estimator of γ.

It is easy to verify that

$$\frac{\alpha(f_n) - \alpha(f)}{\sigma(f)} = \left(\frac{\hat{\gamma}_n}{\gamma} - 1\right) \Delta^{-1}(g)$$

where $\Delta(g) = \sigma(g)/\alpha(g)$.

In view of Theorem 2.3 $\sqrt{n}((\hat{\gamma}_n/\gamma) - 1)\Delta^{-1}(g)$ is distributed asymptotically normal (0,1). This assertion allows us to construct for γ an approximate confidence interval $I_\beta(n)$ with confidence coefficient $1 - \beta$, $0 < \beta < 1$,

$$I_\beta(n) = \left[\frac{\hat{\gamma}_n}{1 + n^{-1/2}\lambda_\beta \Delta(g)}, \quad \frac{\hat{\gamma}_n}{1 - n^{-1/2}\lambda_\beta \Delta(g)} \right],$$

$$\Phi(\lambda_\beta) = 1 - \frac{\beta}{2},$$

and also a test for testing the hypothesis $H_0:\gamma = \gamma_0$.

In particular, let $f(x)$ be a Cauchy density with scale parameter γ. In this case

$$g(x) = \frac{1}{\pi(1 + x^2)}, \quad \hat{\gamma}_n = 2\pi\alpha(f_n), \quad \alpha(g) = \frac{1}{2\pi}, \quad \sigma^2(g) = \frac{1}{2\pi^2}$$

and $\Delta(g) = \sqrt{2}$. The confidence interval for γ is of the form

$$I_\beta(n) = \left[\frac{2\pi\alpha(f_n)}{1 + \sqrt{2/n}\,\lambda_\beta}, \quad \frac{2\pi\alpha(f_n)}{1 - \sqrt{2/n}\,\lambda_\beta} \right].$$

It is well known that it is difficult to estimate the scale parameter of the Cauchy distribution using the maximum likelihood method, while it is impossible to apply the method of moments. From the result obtained one could conclude that our method of constructing an estimator for the scale parameter γ is a simple one and has the same limit distribution as the estimators obtained using the maximum likelihood method.

3. Convergence in Variation of Kernel-Type Density Estimators and its Application to a Non-parametric Estimator of Bayesian Risk in a Classification Problem

Denote by $P\{\cdot\}$ a probability measure on a σ-algebra B_p of Borel sets from a p-dimensional Euclidean space R_p possessing the density $f(\vec{x})$, $\vec{x} = (x_1, x_2, \ldots, x_p)$ with respect to the Lebesgue measure. Let a sample X_i, $i = \overline{1, n}$, be given from a population X distributed with density $f(\vec{x})$. Consider a nonparametric estimator $f_n(\vec{x})$ of density $f(\vec{x})$ of the type (2.2.1). Denote by $P_n\{\cdot\}$ the probability measure corresponding to $f_n(\vec{x})$.

The variation of deviation of the measure $P_n\{\cdot\}$ from the true probability measure $P\{\cdot\}$ will be used as a measure of error of the estimator $P_n\{\cdot\}$. Namely:

$$\rho_1(f_n, f) = \| P_n - P \| = \sup_{A \in B_p} |P_n\{A\} - P\{A\}| =$$

$$= \frac{1}{2} \int | f_n(\vec{x}) - f(\vec{x})| \, d\vec{x}.$$

We pose the following basic problem: Is the convergence in variation of the sequence of measures $P_n\{\cdot\}$, $n = 1, 2, \ldots$ to the measure $\vec{P}\{\cdot\}$ valid without any assumption on the density $f(\vec{x})$? In other words, it is required to prove the correctness - in the sense of convergence in variation - of statistical estimation of an unknown absolutely continous distribution.

The following theorem is valid.

THEOREM 3.1. Let $K(\vec{x})$ be a bounded probability density and the Dirichlet series $\sum_{n=1}^{\infty} ne^{-\gamma\lambda_n}$, $\lambda_n = na_n^{-2p}$ converge for any $\gamma > 0$ (convergence of this series is equivalent to $\log n/\lambda_n \to 0$). Then for any density $f(\vec{x})$:

$$\rho_1(f_n, f) \to 0 \quad \text{a.s.}$$

as $n \to \infty$.

Denote $T_n f = a_n^p \int K(a_n(\vec{x} - \vec{u}))f(\vec{u})\, d\vec{u}$.

LEMMA 3.1. $\|T_n f - f\|_1 \to 0$ as $n \to \infty$ ($\|\cdot\|_1$ is the norm of the space L_1).

Proof. Since the class of continuous and bounded functions which belong to $L_1(R_p)$ is everywhere dense in L_1 there exists for $f(\vec{x})$ and for any $\varepsilon > 0$ a continuous function $f_0(x)$ in L_1 such that

$$\| f_0 - f \|_1 < \varepsilon .$$

Therefore

$$\| T_n f_0 - T_n f \|_1 \leq \| K \|_1 \, \|f_0 - f \|_1 = \|f_0 - f \|_1 < \varepsilon.$$

Hence

$$\| T_n f - f \|_1 \leq \|T_n f_0 - T_n f \|_1 + \|T_n f_0 - f \|_1 < 2\varepsilon +$$

$$+ \| T_n f_0 - f_0 \|_1.$$

Since $f_0(\vec{x})$ is continuous and bounded, $T_n f_0 \to f_0(\vec{x})$ as $n \to \infty$ [33]. Assume that $f_0(\vec{x}) \geq 0$. The general case follows from the representation $f_0(\vec{x}) = f_0^+(\vec{x}) + f_0^-(\vec{x})$ where $f_0^+(\vec{x})$ and $f_0^-(\vec{x})$ are respectively the positive and negative parts of $f_0(\vec{x})$.

It is easy to verify that

$$\int T_n f_0 \, d\vec{x} = \int f_0(\vec{x})\, d\vec{x}.$$

Then in view of Sheffe's theorem [39] $\| T_n f_0 - f_0\|_1$ as $n \to \infty$.
The lemma is thus proved.

Proof of Theorem 3.1. Denote

$$W_n(\vec{x}) = a_n^p \int_{B_x} K(a_n \vec{u})\, d\vec{u}, \qquad B_x = \prod_{i=1}^{p} (-\infty, x_i).$$

Let $F(\vec{x})$ be the distribution function of the random variable X. It is easy to verify that $f_n(\vec{x})$ and $T_n f$ are the probability densities of distributions $F_n(\vec{x}) * W_n(\vec{x})$ and $F(\vec{x}) * W_n(\vec{x})$ respectively ($f_n(\vec{x})$ is the sample distribution function and * denotes the convolution operation).

It is known that $F_n(\vec{x})$ converges to $F(\vec{x})$ a.s. uniformly in $\vec{x} \in R_p$, $W_n(\vec{x})$ converges to a degenerate distribution at 0. Hence $F_n(\vec{x}) * W_n(\vec{x})$ converge to $F(\vec{x})$ a.s. Furthermore for any $\varepsilon > 0$ there exists in R_p a finite interval I_p such that $P(I_p) < \varepsilon/4$, $\bar{I}_p = R_p \backslash I_p$. Then in view of the convergence of $F_n(\vec{x}) * W_n(\vec{x})$ and that of $F(\vec{x}) * W_n(\vec{x})$ to $F(\vec{x})$ there exists a $N(\omega)$ such that for all $n > N(\omega)$

$$A_n^{(1)} = \int_{\bar{I}_p} f_n(\vec{x})\, d\vec{x} < \frac{\varepsilon}{2}, \quad A_n^{(2)} = \int_{\bar{I}_p} Ef_n(\vec{x})\, d\vec{x} < \frac{\varepsilon}{2}. \tag{3.1}$$

It follows from Lemma 2.2.2 and the Schwarz inequality that

$$A_n^{(3)} = \int_{I_p} |f_n(\vec{x}) - Ef_n(\vec{x})|\, d\vec{x} \to 0 \quad \text{a.s.} \tag{3.2}$$

From relations (3.1), (3.2) and the inequality

$$\rho_1(f_n, f) = ||P_n - P|| \leq \frac{1}{2}(A_n^{(1)} + A_n^{(2)} + A_n^{(3)} + A_n^{(4)})$$

follows the assertion of the theorem. Here $A_n^{(4)} = ||T_n f - f||_1 \to 0$ in accordance with Lemma 3.1.

THEOREM 3.2. Let $p = 1$ and $K(x)$ be a function of bounded variation. If the series $\sum_{n=1}^{\infty} e^{-\gamma\lambda_n}$, $\lambda_n = na_n^{-2}$ converges for any $\gamma > 0$, then as $n \to \infty$

$$\rho_1(f_n, f) = ||P_n - P|| \to 0 \quad \text{a.s.}$$

Proof. Let τ be a positive number and $A(\tau) = (-\infty, -\tau] \cup [\tau, \infty)$. We have

$$\begin{aligned}\int |f_n(x) - T_n f|\, dx &= \int_{|x|\leq\tau} |f_n(x) - T_n f|\, dx + \int_{A(\tau)} |f_n(x) - T_n f|\, dx \leq 2\tau \sup_{x \in R_1} |f_n(x) - T_n f| + \int_{|x|\leq\tau} |f_n(x) - \\ &- T_n f|\, dx + 2\int_{A(\tau)} T_n f\, dx \leq 4\tau \sup_{x \in R_1} |f_n(x) - T_n f| + 2\int_{A(\tau)} T_n f\, dx.\end{aligned} \tag{3.3}$$

In view of (2.1.5) the first summand on the r.h.s. of (3.3) converges to zero a.s. Let $\varepsilon > 0$ be arbitrary. Select τ and N so large that

$$\int_{A(\tau)} f(x)\,dx < \frac{\varepsilon}{2} \text{ and } \|T_n f - f\|_1 < \frac{\varepsilon}{2} \quad \text{for} \quad n > N.$$

From here we obtain

$$\int_{A(\tau)} T_n f\,dx < \varepsilon \quad \text{for } n > N. \tag{3.4}$$

Here, in accordance with (3.3) and (3.4)

$$\|f_n - T_n f\|_1 \to 0 \quad \text{a.s.} \tag{3.5}$$

Finally (3.5) and Lemma 3.1 yields the assertion of the theorem.

Remark 1. Theorems 3.1 and 3.2 prove a refinement of the assertion stated by N. N. Chenčov in his lecture [11].

2. As it is well known, the classification problem can be stated as follows: an observation on a random variable X with values in a Euclidean p-dimensional space R_p is available. It is known that the distribution of X is either P_1 or P_2 where P_1 and P_2 are given a priori and it is required to stipulate (classify) to which one of the distributions P_1 of P_2 the observation actually belongs. This classification problem becomes substantially more complex if the prior distributions are unknown and only independent observations on the random variables $X^{(1)}$ and $X^{(2)}$ possessing the distributions P_1 and P_2 respectively are given. Quite a few papers are devoted to classification problems of the latter type (see, e.g. the bibliography in [40]).

In this subsection the strong consistency of Bayes risk estimators is investigated. The empirical classification region constructed herein approaches asymptotically in a certain metric to the unknown optimal set.

Thus, let two probability measures P_1 and P_2 possessing correspondingly densities $f_1(\vec{x})$ and $f_2(\vec{x})$ with respect to the Lebesgue measure be defined on the σ-algebra of B_p-Borel sets in R_p.

A Bayesian classification rule $\delta(X)$ is defined as follows: an observation X belongs to the population Γ_1 or Γ_2 with distributions P_1 or P_2 respectively depending on whether

$$X \in E_0 = \{\vec{x}: f_1(\vec{x}) - Cf_2(\vec{x}) \geq 0\} \quad \text{or} \quad X \in \overline{E_0} = R_p \setminus E_0.$$

where $C = C_2(1-\pi)/C_1\pi$. Here C_1 and C_2 are the costs of erroneous classifications and π and $1-\pi$ are the prior probabilities associated with Γ_1 and Γ_2. It is assumed that C is known.

In the class of all subdivisions of the space R_p into two disjoint sets Ω_1 and Ω_2, $\Omega_1 \cup \Omega_2 = R_p$ the expected loss (risk) of an incorrect classification is equal to

$$R(\chi_{\Omega_1}) = \int [C_1\pi(1-\chi_{\Omega_1}(\vec{x}))f_1(\vec{x}) + C_2(1-\pi)\chi_{\Omega_1}(\vec{x})f_2(\vec{x})]\,d\vec{x},$$

where χ_Ω is the indicator of the set Ω.

A Bayesian risk is defined as $R^* = \min_{\Omega \in B_p} R(\chi_\Omega) = R(\chi_{E_0})$. Now, let $f_1(\vec{x})$ and $f_2(\vec{x})$ be unknown: then a Bayesian decision rule usually cannot be defined and hence one cannot evaluate analytically the Bayesian risk. However, if sufficiently correctly classified data (the so-called training samples) of sizes n_i from Γ_i, $i = 1, 2$, are available one can then empirically determine the class of nonparametric decision rules $\delta_n(X)$:

$$X \in \Gamma_1 \quad \text{if} \quad X \in E^n = \{\vec{x}: f_{n_1}(\vec{x}) \geq C f_{n_2}(\vec{x}), \quad \text{and}$$

$$X \in \Gamma_2 \quad \text{if} \quad X \in R_p \backslash E^{(n)}$$

where $n = (n_1, n_2)$ and $f_{ni}(\vec{x})$ are estimators of the density $f_i(\vec{x})$ of the type (4.1). The risk corresponding to the rule $\delta_n(X)$ will be denoted by $R^{(n)} = R(\chi_{E^{(n)}})$.

The difference of the risks $R^{(n)}$ and R^* satisfies the following inequality [41]:

$$0 \leq R^{(n)} - R^* \leq 2C_1\pi \,||P_{n_1} - P_1|| + 2C_2(1 - \pi)\,||P_{n_2} - P_2||, \tag{3.6}$$

where P_{n_i} is the measure corresponding to $f_{n_i}(\vec{x})$.

The result - stated below as a theorem - follows from Theorem 3.1 or Theorem 3.2.

THEOREM 3.3. If a_i, $i = 1, 2$, and $K(\vec{x})$ satisfy the conditions of Theorems 3.1 or 3.2, then the risk $R^{(n)}$ corresponding to the decision rule $\delta_n(X)$ converges a.s. to R^* as $n_1, n_2 \to \infty$:

$$R^{(n)} \to R^* \quad \text{a.s.} \tag{3.7}$$

We shall use the distance $d(E_0 \Delta E^{(n)}) = \max\,[P_1(E^{(n)}\Delta E_0), P_2(E^{(n)}\Delta E_0)]$ as the measure of deviation between E_0 and $E^{(n)}$. Here $E^{(n)}\Delta E_0$ is the symmetric difference of $E^{(n)}$ and E_0. Denote by $B_0 = \{\vec{x}: C_1\pi f_1(\vec{x}) - C_2(1 - \pi)f_2(\vec{x}) = 0\}$ and assume that $P_i(B_0) = 0$, $i = 1, 2$. It is easy to verify that

$$0 \leq R^{(n)} - R^* = \int [C_1\pi f_1(\vec{x}) - C_2(1 - \pi)f_2(x)][\chi_{E_0}(\vec{x}) - \chi_{E^{(n)}}(\vec{x})]\, d\vec{x} = \int_{E^{(n)}\Delta E_0} |C_1\pi f_1(\vec{x}) - C_2(1 - \pi)f_2(\vec{x})|\, d\vec{x}. \tag{3.8}$$

The measure $P = P_1 + P_2$ $(P_i(A) = \int_A f_i(\vec{x})\, d\vec{x}$, $A \in B_p)$ is absolutely continuous with respect to the measure

$$\mu(A) = \int_A |C_1 \pi f_1(\vec{x}) - C_2(1 - \pi) f_2(\vec{x})| \, d\vec{x}.$$

Indeed, let $A \in B_p$, $A \cap B_0 = \phi$ and $\mu(A) = 0$ then $\ell(A) = 0$ (ℓ is the Lebesgue measure); hence $P_i(A) = 0$ i.e. $P(A) = 0$. Finally let $B_0 \cap A \neq \phi$ and $\mu(A) = 0$ then $\ell(A - B_0) = 0$ and hence $P(A) = 0$. Thus $P << \mu$. From this result and (3.7) and (3.8) follows

THEOREM 3.4. If a_i, $i = 1, 2$ and $K(\vec{x})$ satisfy the conditions of Theorem 3.3 then $P(E^{(n)} \Delta E_0) \to 0$ a.s. as $n_1, n_2 \to \infty$ and thus $d(E^{(n)} \Delta E_0) \to 0$ a.s. as $n_1, n_2 \to \infty$.

THEOREM 3.5. Let $K(\vec{x})$, $\vec{x} \in R_p$, be a bounded finite probability density. If $na_n^{-p} \to \infty$ as $n \to \infty$ then for any density $f(\vec{x})$

$$\|P_n - P\| = \sup_{A \in B_p} |P_n\{A\} - P\{A\}| \to 0$$

in probability.

For convenience we shall prove a lemma which is apparently known. However, the author is not aware where a proof of this lemma is presented. It is a corollary of Lebesgue's theorem on bounded convergence and Fatou's lemma.

LEMMA 3.2. Let $\{\phi_n^{(i)}(\vec{x})\}$, $\vec{x} \in R_p$, $i = 1, 2, 3$, be sequences of measurable functions such that

$$\phi_n^{(1)}(\vec{x}) \leq \phi_n^{(2)}(\vec{x}) \leq \phi_n^{(3)}(\vec{x}), \quad n = 1, 2, \ldots$$

and $\lim_{n\to\infty} \phi_n^{(i)}(\vec{x}) = \phi^{(i)}(\vec{x})$, $i = 1, 2, 3$, a.e. Moreover, let

$$\int \phi_n^{(i)}(\vec{x}) \, d\vec{x} \to \int \phi^{(i)}(\vec{x}) \, d\vec{x} \quad i = 1 \quad \text{and} \quad i = 3$$

as $n \to \infty$. Then

$$\int \phi_n^{(2)}(\vec{x}) \, d\vec{x} \to \int \phi^{(2)}(\vec{x}) \, d\vec{x} \quad \text{as} \quad n \to \infty.$$

<u>Proof</u>. Since $0 \leq \phi_n^{(2)}(\vec{x}) - \phi_n^{(1)}(\vec{x}) \to \phi^{(2)}(\vec{x}) - \phi^{(1)}(\vec{x})$ and $0 \leq \phi_n^{(3)}(\vec{x}) - \phi_n^{(2)}(\vec{x}) \to \phi^{(3)}(\vec{x}) - \phi^{(2)}(\vec{x})$, we have in view of Fatou's lemma that

$$\int \phi^{(3)}(\vec{x}) \, d\vec{x} - \int \phi^{(2)}(\vec{x}) \, d\vec{x} = \int \lim_{n\to\infty} (\phi_n^{(3)}(\vec{x}) - \phi_n^{(2)}(\vec{x})) \, d\vec{x} \leq$$

$$\leq \underline{\lim}_{n\to\infty} \int (\phi_n^{(3)}(\vec{x}) - \phi_n^{(2)}(\vec{x})) \, d\vec{x} = \int \phi^{(3)}(\vec{x}) \, d\vec{x} - \overline{\lim}_{n\to\infty} \int \phi_n^{(2)}(\vec{x}) \, d\vec{x} \tag{3.9}$$

and

$$\int \phi^{(2)}(\vec{x})\, d\vec{x} - \int \phi^{(1)}(\vec{x})\, d\vec{x} = \int \lim_{n\to\infty} (\phi_n^{(2)}(\vec{x}) - \phi_n^{(1)}(\vec{x}))\, d\vec{x} \leq$$

$$\leq \underline{\lim}_{n\to\infty} \int (\phi_n^{(2)}(\vec{x}) - \phi_n^{(1)}(\vec{x}))\, d\vec{x} = \underline{\lim}_{n\to\infty} \int \phi_n^{(2)}(\vec{x})\, d\vec{x} - \int \phi^{(1)}(\vec{x})\, d\vec{x}. \tag{3.10}$$

(Here the following simple assertion is used: Let $\{x_n\}$ and $\{y_n\}$ be two numerical sequences such that $\{y_n\}$ is convergent, then $\underline{\lim}_{n\to\infty} (x_n \pm y_n) = \underline{\lim}_{n\to\infty} x_n \pm \lim_{n\to\infty} y_n$).

From (3.9) and (3.10) the inequalities

$$\overline{\lim}_{n\to\infty} \int \phi_n^{(2)}(\vec{x})\, d\vec{x} \leq \int \phi^{(2)}(\vec{x})\, d\vec{x} \leq \underline{\lim}_{n\to\infty} \int \phi_n^{(2)}(\vec{x})\, d\vec{x}$$

follow. On the other hand $\underline{\lim}_{n\to\infty} \int \phi_n^{(2)}(\vec{x})\, d\vec{x} \leq \overline{\lim}_{n\to\infty} \int \phi_n^{(2)}(\vec{x})\, d\vec{x}$.

The lemma is thus proved.

LEMMA 3.3. Let $K(\vec{x})$ satisfy the conditions of Theorem 3.5. Then as $n \to \infty$

$$a_n^p \int K^m(a_n(\vec{x} - \vec{y}))f(\vec{y})\, d\vec{y} \to f(\vec{x}) \int K^m(\vec{y})\, d\vec{y}, \quad m \geq 1,$$

for allmost all $\vec{x} \in R_p$.

<u>Proof</u>. We have

$$|a_n^p \int K^m(a_n(\vec{x} - \vec{y}))f(\vec{y})\, d\vec{y} - f(\vec{x}) \int K^m(\vec{y})\, d\vec{y}| \leq a_n^p \int_{S(x,\delta/a_n)} K^m \times$$

$$\times (a_n(\vec{x} - \vec{y}))|f(\vec{y}) - f(\vec{x})|\, d\vec{y} + a_n^p \int_{\bar{S}(x,\delta/a_n)} K^m \times$$

$$\times (a_n(\vec{x} - \vec{y}))|f(\vec{y}) - f(\vec{x})|\, d\vec{y} = L_n^{(1)} + L_n^{(2)},$$

where $S(\vec{x}), \delta/a_n)$ is a sphere of radius δ/a_n with the center at point $\vec{x}$ and $\bar{S}(\vec{x}, \delta/a_n) = R_p \backslash S(\vec{x}, \delta/a_n)$.

Since $K(\vec{y})$ is finite there exists a $\delta > 0$ such that $K(\vec{y}) \equiv 0$ for all $\vec{y} \in \bar{S}(0, \delta)$. It thus follows from here that for all $n \geq 1$

$$L_n^{(2)} = \int_{\bar{S}(0,\delta)} K^m(\vec{y})|f(\vec{x} - a_n^{-1}\vec{y}) - f(\vec{x})|\, d\vec{y} = 0.$$

Denote by F the set of Lebesgue points of the function $f(\vec{x})$. Then for all points $\vec{x} \in F$

$$L_n^{(1)} \leq \sup_{\vec{y} \in R_p} K^m(y)\delta^p(\delta/a_n)^{-p} \int_{S(\vec{x},\delta/a_n)} |f(\vec{y}) - f(\vec{x})|\, d\vec{y} \to 0$$

as $n \to \infty$.

The lemma is proved.

Proof of Theorem 3.5. We have

$$(E|f_n(\vec{x}) - f(\vec{x})|)^2 \leq E(f_n(\vec{x}) - f(\vec{x}))^2.$$

However,

$$E(f_n(\vec{x}) - f(\vec{x}))^2 = Df_n(\vec{x}) + b(f_n),$$

where

$$b(f_n) = Ef_n(\vec{x}) - f(\vec{x}) \to 0 \quad \text{and } na_n^{-p}\, Df_n(\vec{x}) \leq a_n^p \int K^2 \times$$

$$\times\, (a_n(\vec{x} - \vec{u}))f(\vec{u})\, d\vec{u} \to f(\vec{x}) \int K^2(\vec{u})\, d\vec{u} \quad \text{as} \quad n \to \infty$$

for almost all $\vec{x} \in R_p$ in view of Lemma 3.3. Hence $Df_n(\vec{x}) \to 0$ as $n \to \infty$ a.e.

Thus for almost all $\vec{x} \in R_p$

$$E|f_n(\vec{x}) - f(\vec{x})| \to 0 \quad \text{as} \quad n \to \infty. \tag{3.11}$$

Next utilizing the Fubini theorem we have

$$E||P_n - P|| = \frac{1}{2} \int E|f_n(\vec{x}) - f(\vec{x})|\, d\vec{x}.$$

Set

$$\phi_n^{(1)}(\vec{x}) \equiv 0, \quad \phi_n^{(2)}(\vec{x}) = E|f_n(\vec{x}) - f(\vec{x})|,$$

$$\phi_n^{(3)}(\vec{x}) = Ef_n(\vec{x}) + f(\vec{x}).$$

Clearly $\phi_n^{(1)}(\vec{x}) \leq \phi_n^{(2)}(\vec{x}) \leq \phi_n^{(3)}(\vec{x})$ and moreover, in view of (3.11), $\phi_n^{(2)}(\vec{x}) \to 0$ as $n \to \infty$. Also $\phi_n^{(3)}(\vec{x}) \to 2f(\vec{x})$ and $\int \phi_n^{(3)}(\vec{x})\, d\vec{x} \to 2 \int f(\vec{x})\, d\vec{x} = 2$. Therefore in accordance with Lemma 3.2 we obtain

$$\int \phi_n^{(2)}(\vec{x})\, d\vec{x} \to 0 \quad \text{as} \quad n \to \infty.$$

The theorem is thus proved.

Chapter 3

LIMITING DISTRIBUTIONS OF DEVIATIONS OF KERNEL-TYPE DENSITY ESTIMATORS

1. Limiting Distribution of Maximal Deviation of Kernel-Type Estimators

1. In the preceding chapter we investigated the problem of strong consistency in various functional metrics of kernel-type density estimators. The theorems proved therein were basically of a qualitative nature. The problem of accuracy and reliability of approximations for large samples was not addressed in these theorems. In this section the limiting behavior of the distribution of the maximal deviation of an empirical density $f_n(x)$ of the type (1.1.1) from the unknown density $f(x)$ (the so-called N. V. Smirnov problem) is investigated. The theorems we are going to prove permit us to carry out statistical estimation of the degree of approximation and to construct for an unknown theoretical density a confidence region with a preassigned confidence coefficient. As it is known, the groundbreaking work in this direction is N. V. Smirnov's paper [2] in which the limiting distribution of the maximal deviation for the case when $f_n(x)$ is a 'histogram' was obtained.

The problem of obtaining a limiting distribution of the maximal deviation of estimator (1.1.1) on a finite interval $[a, b]$ was apparently solved for the first time by us in [74] under the assumption that $f(x) \geq \mu > 0$, $x \in [a, b]$ (the regular case). In addition to this problem the following interesting problem arises: is it possible to drop the condition $f(x) \geq \mu > 0$, $x \in [a, b]$ and consider the 'singular' case in the sense that the density $f(x)$ may vanish at certain isolated points of the interval or approach zero at infinity in an unbounded manner. In this section a solution to these problems is presented.

2. Assumption and Notation

We shall use the following assumptions concerning $f(x)$ and $K(x)$.

1°. $f(x)$ is continuous and bounded;

2°. $f(x)$ possesses a bounded derivative;

3°. $f(x)$ possesses a second order bounded derivative;

4°. $K(x)$ vanishes outside of a finite interval $[-A, A]$, is bounded, symmetric and $\int K(u)\,du = 1$.

Let $f_n(x)$ denote the estimator (1.1.1):

$$f_n(x) = a_n n^{-1} \sum_{i=1}^{n} K(a_n(x - X_i)).$$

Let $E_1, E_2, \ldots, E_s$ be finite and disjoint intervals in $R = (-\infty, \infty)$ and $E_{s+1} = R \setminus \bigcup_{k=1}^{s} E_k$. The number of intervals $s = s_n$ increases with the increase of the sample size n. Denote the length of the interval E_k

by h_k and the sum of lengths of these intervals by Δ_{s_n}. Assume that the choice of the intervals E_k, $k = \overline{1, s_n}$ is subject to the conditions: $2Aa_n^{-1} \leq h_k \leq C_0 a_n^{-1}$ (as above C_0, C_1, ... denote absolute constants). In view of the assumption Δ_{s_n} is finite for a given n but obviously may increase with n.

Next, let $t_1, t_2, \ldots t_{s_n}$ be the centers of the intervals $E_1, E_2, \ldots, E_{s_n}$ and set

$$\xi_n(t_j) = n^{-1/2} \sum_{i=1}^{n} \xi_n^{(i)}(t_j), \quad I \leq j \leq s_n,$$

where

$$\xi_n^{(i)}(t_j) = [K(a_n(t_j - X_i)) - a_n^{-1} Ef_n(t_j)][\int K^2(a_n(t_j - u))f \times$$

$$\times (u)\, du - a_n^{-2}(Ef_n(t_j))^2]^{-1/2},$$

$$\varepsilon_n = \max_{\substack{1\leq i,j\leq s_n \\ i\neq j}} |E(\xi_n(t_j)\xi_n(t_i))|, \quad \mu_{s_n} = \min_{x \in R \setminus E_{s_n+1}} f(x) > 0.$$

3. Lemma on Probabilities of Large Deviations

The derivation of the limiting distribution of statistics which are utilized for construction of a confidence region for f(x) is based on a number of auxiliary propositions. The basic one is Lemma 1.3 on probabilities of large deviations which is also of interest on its own.

LEMMA 1.1. Under our assumptions 1° and 4° on f(x) and K(x) we have

$$\varepsilon_n \leq C_1 \frac{\Delta_{s_n}}{s_n \mu_{s_n}}.$$

as $\Delta_{s_n}/s_n\mu_{s_n} \to 0$.

Proof. We note that $K(2A - x) \equiv 0$ for $x \in [-A, A]$ and $K(a_n(t_j - u)) \equiv 0$ for $u \in E_{s_n+1}$. The first identity is obvious while the second follows from the inequality $\min_{1\leq j\leq s_n} |t_j - u| \geq Aa_n^{-1}$. Therefore we have

$$\varepsilon_n \leq C_2 a_n^{-2} \max_{\substack{1\leq i,j\leq s_n \\ i\neq j}} [a_n^{-1}\mu_{s_n} \int K^2(u)\, du - a_n^{-2}(E(f_n(t_j))^2]^{-1/2} \times$$

$$\times [a_n^{-1}\mu_{s_n} \int K^2(u)\, du - a_n^{-2}(Ef_n(t_i))^2]^{-1/2}. \tag{1.1}$$

Next, some elementary calculations show that

$$C_0^{-1} s_n^{-1} \Delta_{s_n} \leq a_n^{-1} \leq (2A)^{-1} s_n^{-1} \Delta_{s_n}. \tag{1.2}$$

Utilizing inequalities (1.2) in expression (1.1) and setting $M = \max_{x \in R} f(x)$ we obtain

$$\varepsilon_n \leq C_2 (2A)^{-2} \frac{\Delta_{s_n}}{s_n \mu_{s_n}} \left[C_0^{-1} \int K^2(u)\, du - M^2 (2A)^{-2} \frac{\Delta_{s_n}}{s_n \mu_{s_n}} \right]^{-1}. \tag{1.3}$$

The difference in (1.3) for n sufficiently large is evidently positive and converges to a finite limit. Thus the assertion of the lemma is verified.

LEMMA 1.2. Under the conditions of Lemma 1.1

$$\max_{1 \leq i \leq n} \max_{1 \leq j \leq s_n} |\xi_n^{(i)}(t_j)| \leq C_3 \left(\frac{\mu_{s_n} \Delta_{s_n}}{s_n} \right)^{-1/2} \quad \text{a.s.}$$

The proof of Lemma 1.2 is analogous to the proof of Lemma 1.1.

LEMMA 1.3. Let f(x) and K(x) satisfy conditions 1° and 3° respectively. Let λ_{s_n} be a non-negative and increasing function of s_n. If as s_n increases

$$s_n^{-1} \Delta_{s_n} \mu_{s_n}^{-1} \lambda_{s_n}^2 \to 0 \quad \text{and} \quad \frac{\sqrt{s_n}\, \lambda_{s_n}^m}{\sqrt{n \mu_{s_n} \Delta_{s_n}}} \to 0,$$

then

$$P_n = P\{|\xi_n(t_j)| \geq \lambda_{s_n}, \quad j = \overline{1, r}\} = \left(\frac{2}{\sqrt{2\pi}} \int_{\lambda_{s_n}}^{\infty} \exp\left(-\frac{x^2}{2} \right) dx \right)^r \times$$

$$\times \left(1 + O\left(\frac{\lambda_{s_n}^2 \cdot \Delta_{s_n}}{s_n \mu_{s_n}} \right) + O\left(\frac{\sqrt{s_n}\, \lambda_{s_n}^m}{\sqrt{n \mu_{s_n} \Delta_{s_n}}} \right) \right),$$

where $m = \max(3, r)$.

Remark 1. The proof of Lemma 1.3 will remain essentially the same if in place of the first r normalized deviations $\xi_n(t_j)$, $j = \overline{1, r}$ we take arbitrary r normalized deviations $\xi_n(t_{k_j})$, $j = \overline{1, r}$ ($k_1, k_2, \ldots, k_r$ are arbitrary combinations of r numbers from the sequences $1, 2, \ldots, s_n$). The first r normalized deviations are used only to simplify the notation.

The proof of this lemma is based on an application of conjugate distributions introduced by H. Cramér [42] and is generalized to the multivariate case in a natural manner.

Consider a sequence of independent r-dimensional random variables

$$\xi_i = (\xi_n^{(i)}(t_1), \ldots, \xi_n^{(i)}(t_r)), \quad i = 1, 2, \ldots, n,$$

possessing the same distribution function $V_n(\vec{x})$, $\vec{x} = (x_1, x_2, \ldots x_r)$.

Denote by $W_n(\vec{x})$, $\vec{x} \in R_r$, the distribution function of the sum

$\xi_1 + \xi_2 + \ldots + \xi_n$ and by $F_n(\vec{x})$ the distribution function of the normalized sum $n^{-1/2}(\xi_1 + \ldots + \xi_n)$.

Thus we have

$$F_n(\vec{x}) = W_n(\sqrt{n}\,\vec{x}). \tag{1.4}$$

Let $\bar{V}_n(\vec{x})$ be a distribution function conjugated to $V_n(\vec{x})$ and be defined by the inequality

$$\bar{V}_n(\vec{x}) = R_n^{-1} \int_{-\infty}^{x} e^{(\tau,\vec{y})}\, dV_n(\vec{y}), \qquad \|\tau\| \le \tau_0$$

where $R_n = \int e^{(\tau,\vec{y})}\, dV_n(\vec{y})$ and $(\tau, \vec{y})$ is the scalar product of the vectors $\tau, \vec{y} \in R_r$.

Consider the sequence $\bar{\xi}_k = (\bar{\xi}_{k1}, \ldots, \bar{\xi}_{kr})$, $k = \overline{1, n}$, of independent random vectors possessing the same distribution function $\bar{V}_n(x)$ with the mean vector $E\bar{\xi}_1 = \bar{m}_n = (m_1, \ldots, m_r)$ and the covariance matrix $\{\bar{R}_{ij}\} = E(\bar{\xi}_1 \cdot \bar{\xi}_1')$.

Denote by $\bar{W}_n(\vec{x})$ the distribution function of the sum $\bar{\xi}_1 + \ldots + \bar{\xi}_n$ and by $\bar{F}_n(x)$ the distribution function of the normalized sum $n^{-1/2}(\bar{\xi}_1 + \bar{\xi}_2 + \ldots + \bar{\xi}_n - n \cdot \bar{m}_n)$. Clearly the following relation is valid:

$$\bar{F}_n(\vec{x}) = \bar{W}_n(\vec{x} \cdot \sqrt{n} + n \cdot \bar{m}_n). \tag{1.5}$$

Arguing as in Cramér's paper [42] it is easy to derive the following relationship between $W_n(\vec{x})$ and $\bar{W}_n(\vec{x})$:

$$W_n(\vec{x}) = (R_n)^n \int_{-\infty}^{\vec{x}} e^{-(\tau,\vec{y})}\, d\bar{W}_n(\vec{y}). \tag{1.6}$$

Substituting functions $F_n(\vec{x})$ and $\bar{F}_n(\vec{x})$ defined by formulas (1.4) and (1.5) into (1.6) we obtain

$$F_n(\vec{x}) = (R_n)^n \exp\{-n(\tau, \bar{m}_n)\} \int_{-\infty}^{x-\sqrt{n}\cdot\bar{m}_n} \exp\{-(\tau, \vec{y})\sqrt{n}\}\, d\bar{F}_n(\vec{y}). \tag{1.7}$$

Equation (1.7) easily implies that

$$P_n' = P\left\{n^{-1/2} \sum_{i=1}^{n} \xi_n^{(i)}(t_k) \ge \lambda_{s_n}, \quad k = \overline{1, r}\right\} = \int_{T_{s_n}}^{\infty} dF_n(\vec{x}) =$$

$$= (R_n)^n \exp\{-n(\tau, \bar{m}_n)\} \int_{H_n}^{\infty} \exp\{-(\tau, \vec{y})\sqrt{n}\}\, d\bar{F}_n(\vec{y}), \tag{1.8}$$

where

$$T_{s_n} = (\lambda_{s_n}, \ldots, \lambda_{s_n}) \quad \text{and} \quad H_n = T_{s_n} - \sqrt{n}\cdot\bar{m}_n.$$

Below we shall always assume that

$$\tau = \tau_n = (n^{-1/2}\lambda_{s_n}, \ldots, n^{-1/2}\lambda_{s_n}), \quad \lambda_{s_n} = O(\sqrt{n}).$$

Our first task is to investigate the behavior of P_n.

We shall show that

$$\sqrt{n}\cdot m_i = \lambda_{s_n} + O\left(\frac{\Delta_{s_n}}{s_n\mu_{s_n}}\lambda_{s_n}\right) + O\left(\frac{\sqrt{s_n}\cdot\lambda_{s_n}^2}{\sqrt{n\mu_{s_n}\Delta_{s_n}}}\right), \quad i = \overline{1, r}, \tag{1.9}$$

and that the bound on the remainder in (1.9) is uniform with respect to i, $1 \leq i \leq r$.

Indeed, we have

$$\begin{aligned}
\sqrt{n}\cdot m_i &= \frac{\sqrt{n}}{R_n}\int x_i e^{(\tau_n, \vec{x})} dV_n(\vec{x}) = \lambda_{s_n} R_n^{-1}\Big[\int x_i x_1\, dV_n(\vec{x}) + \\
&+ \ldots + \int x_i^2\, dV_n(\vec{x}) + \ldots + \int x_i x_r\, dV_n(\vec{x})\Big] + \\
&+ \frac{1}{2}\sqrt{n}\, R_n^{-1}\int x_i(\tau_n, \vec{x})^2 \exp\{\theta_i(\tau_n, \vec{x})\}\, dV_n(\vec{x}) = \\
&= \lambda_{s_n} R_n^{-1}(R_{1i} + \ldots + 1 + \ldots + R_{ir}) + \\
&+ \frac{1}{2} R_n^{-1}\sqrt{n}\int x_i(\tau_n, \vec{x})^2 \exp\{\theta_i(\tau_n, \vec{x})\}\, dV_n(\vec{x}).
\end{aligned} \tag{1.10}$$

where

$$0 \leq \Theta_i < 1, \quad R_{ij} = E(\xi_n(t_i)\xi_n(t_j)).$$

By virtue of Lemma 1.2 we have

$$\begin{aligned}
&\sqrt{n}\,\Big|\int x_i(\tau_n, \vec{x})^2 \exp\{\Theta_i(\tau_n, \vec{x})\}\, dV_n(\vec{x})\Big| \leq r n^{-1/2}\lambda_{s_n}^2 \int |x_i| \times \\
&\times \sum_{j=1}^{r} x_j^2 \exp\{\Theta_i(\tau_n, \vec{x})\}\, dV_n(\vec{x}) \leq C_4 \frac{\lambda_{s_n}^2\sqrt{s_n}}{\sqrt{n\mu_{s_n}\Delta_{s_n}}} \exp\left\{ rC_3\Theta_i \times \right. \\
&\left. \times \frac{\sqrt{s_n}\lambda_{s_n}}{\sqrt{n\mu_{s_n}\Delta_{s_n}}}\right\} \int \sum_{j=1}^{r} x_j^2\, dV_n(\vec{x}) \leq C_5 \frac{\lambda_{s_n}^2\sqrt{s_n}}{\sqrt{n\mu_{s_n}\Delta_{s_n}}}
\end{aligned} \tag{1.11}$$

and

$$R_n = 1 + O\left(\frac{\lambda_{s_n}^2}{n}\right). \tag{1.12}$$

Finally in view of (1.11) and (1.12), (1.10) yields (1.9).

Arguing in the same manner as in the derivation of (1.9) we obtain

$$\bar{R}_{ij} = R_{ij} + \ell_{ij}, \tag{1.13}$$

where

$$R_{ij} = \int x_i x_j \, dV_n(\vec{x}) \quad \text{and} \quad |\ell_{ij}| \le C_5 \frac{\sqrt{s_n} \cdot \lambda_{s_n}}{\sqrt{n\mu_{s_n} \Delta_{s_n}}}, \quad i, j = \overline{1, r}.$$

Indeed to prove (1.13) it is sufficient to consider the expression

$$R_n^{-1} \int x_i x_j \exp\{(\tau_n, \vec{x})\} \, dV_n(\vec{x}).$$

Evidently the following equality is valid:

$$\int x_i x_j \exp\{(\tau_n, \vec{x})\} \, dV_n(\vec{x}) = \int x_i x_j \, dV_n(\vec{x}) + A_1 + A_2, \tag{1.14}$$

where

$$A_1 = n^{-1/2} \lambda_{s_n} \int x_i x_j (x_1 + \ldots + x_r) \, dV_n(\vec{x})$$

and

$$A_2 = \frac{1}{2} n^{-1} \lambda_{s_n}^2 \int x_i x_j (x_1 + \ldots + x_r)^2 \exp\{\Theta_i (\tau_n, \vec{x})\} \, dV_n(\vec{x}),$$

$$0 < \Theta_i < 1.$$

In view of Lemma 1.2 we have

$$A_1 \le C_3 \cdot r \frac{\lambda_{s_n} \sqrt{s_n}}{\sqrt{n\mu_{s_n} \Delta_{s_n}}}, \qquad A_2 \le C_5 \frac{\lambda_{s_n}^2 \cdot s_n}{n\mu_{s_n} \Delta_{s_n}}. \tag{1.15}$$

Now (1.15), (1.12) and (1.14) yield (1.13).

We shall now investigate the expression appearing on the r.h.s. of (1.8). Here we are going to use V. V. Sazanov's results [43] dealing with the estimation of the rate of convergence in a multidimensional central limit theorem. According to these results the distribution function $\bar{F}_n(\vec{x})$ admits the decomposition

$$\bar{F}_n(\vec{x}) = G_n(\vec{x}) + Q_n(\vec{x}). \tag{1.16}$$

where $G_n(\vec{x})$ is the distribution function of a normal law with the same first and second moments as the distribution function $\bar{V}_n(\vec{x})$ and the function $Q_n(\vec{x})$ admits a uniform bound of the form

$$\sup_{x \in R_r} |Q_n(\vec{x})| \le C_6 \left(\sum_{i=1}^{r} \frac{\Delta_n^{(ii)}}{\Delta_n} \rho_n^{(i)} \right) n^{-1/2}, \tag{1.17}$$

where $\rho_n^{(i)} = (E|\bar{\xi}_{1i} - m_i|^3)(E(\bar{\xi}_{1i} - m_i)^2)^{-3/2}$, Δ_n is the determinant of the matrix of correlation coefficients of the random vector $\overline{\xi}_1$ and Δ_n^{ii} are the corresponding minors of this determinant.

Since, in view of Lemma 1.2 the distribution function $V_n(\vec{x})$ is concentrated on the cube with the sides $O((\mu_{s_n}\Delta_{s_n}/s_n)^{-1/2})$ the distribution function $\overline{V}_n(\vec{x})$ is also concentrated on the same cube. Therefore

$$\rho_n^{(i)} \leq \frac{\max_i |\overline{\xi}_{1i} - m_i|}{\sqrt{D\overline{\xi}_{1i}}} \leq C_7 \left(\frac{\mu_{s_n}\Delta_{s_n}}{s_n}\right)^{-1/2}, \quad i = \overline{1, r}, \tag{1.18}$$

and

$$\sum_{i=1}^{r} \frac{\Delta_n^{(ii)}}{\Delta_n} \to r \quad \text{as} \quad n \to \infty .$$

Thus

$$\sup_{\vec{x} \in R_r} |Q_n(\vec{x})| \leq C_8 \left(\frac{s_n}{n\mu_{s_n}\Delta_{s_n}}\right)^{1/2} . \tag{1.19}$$

We now return to the expression for P_n'. Substituting (1.16) into (1.8) we obtain

$$P_n' = (R_n)^n \exp\{-n(\tau_n, \overline{m}_n)\}\left\{\int_{H_n}^{\infty} \exp\{-(\tau_n, \vec{y})\sqrt{n}\}\, dG_n(\vec{y}) + \right.$$

$$\left. + \int_{H_n}^{\infty} \exp\{-(\tau_n, \vec{y})\sqrt{n}\}\, dQ_n(\vec{y})\right\} = I_1 + I_2. \tag{1.20}$$

First we shall estimate I_2.

Utilizing the formula of integration by parts for a multidimensional integral (see, e.g. [44]) we arrive at

$$\left|\int_{H_n}^{\infty} \exp\{-(\tau_n, \vec{y})\sqrt{n}\}\, dQ_n(\vec{y})\right| \leq C_9 \left(\frac{s_n}{n\mu_{s_n}\Delta_{s_n}}\right)^{1/2} . \tag{1.21}$$

The factor $(R_n)^n \exp[-n(\tau_n, \overline{m}_n)]$ which appears in I_2 can be expressed as $\exp[n(\log R_n - (\tau_n, \overline{m}_n))]$. Next, in view of Lemmas 1.1 and 1.2 it is easy to verify that

$$n[\log R_n - (\tau_n, \overline{m}_n)] = -\frac{r}{2}\lambda_{s_n}^2 + O\left(\frac{\Delta_{s_n}\lambda_{s_n}^2}{s_n \cdot \mu_{s_n}}\right) +$$

$$+ O\left(\frac{\sqrt{s_n}\cdot\lambda_{s_n}^3}{\sqrt{n\mu_{s_n}\cdot\Delta_{s_n}}}\right) . \tag{1.22}$$

Hence in view of (1.21) and (1.22) we arrive at

$$I_2 = O\left(\frac{\sqrt{s_n}}{\sqrt{n\mu_{s_n}\Delta_{s_n}}} \exp\left\{-\frac{r}{2}\lambda_{s_n}^2\right\}\right). \tag{1.23}$$

We now proceed to the calculation of the main term in I_1. Denote

$$L_n(\vec{x}) = (2\pi)^{-r/2}\Delta_n^{-1/2} \exp\left\{-\frac{1}{2\Delta_n}\sum_{i,j}\Delta_n^{(ij)}x_i x_j\right\} - (2\pi)^{-r/2} \times$$
$$\times \exp\left\{-\frac{1}{2}\sum_{k=1}^{r} x_k^2\right\}. \tag{1.24}$$

We have

$$\sum_{i,j}\Delta_n^{(ij)}x_i x_j = \sum_{i=1}^{r} x_i^2 + \sum_{i=1}^{r}\tilde{\ell}_i x_i^2 + \sum_{i=1}^{r}\Sigma'_i) x_i^2 + \sum_{j\neq k}\Delta_{jk}x_j x_k, \tag{1.25}$$

where

$$\Sigma'_i = \Delta_n^{(ii)} - \bar{R}_{11}\bar{R}_{22}\,\dots\,\bar{R}_{i-1i-1}\bar{R}_{i+1i+1}\,\dots\,\bar{R}_{rr};$$

and moreover

$$|\tilde{\ell}_i| \leq C_{10}\frac{\sqrt{s_n}\cdot\lambda_{s_n}}{\sqrt{n\mu_{s_n}\Delta_{s_n}}}, \quad i = 1, 2, \dots r. \tag{1.26}$$

In view of (1.13)

$$\Delta_n^{-1} = 1 + O\left(\frac{\Delta_{s_n}}{s_n\mu_{s_n}}\right) + O\left(\frac{\sqrt{s_n}\cdot\lambda_{s_n}}{\sqrt{n\mu_{s_n}\Delta_{s_n}}}\right). \tag{1.27}$$

Utilizing (1.13), (1.25), (1.26) and (1.27) we conclude that

$$-\frac{1}{2\Delta_n}\sum_{j,k}\Delta_n^{(jk)}x_j x_k = -\frac{1}{2}\sum_{k=1}^{r} x_k^2 + \delta_n(\vec{x}) \tag{1.28}$$

where

$$|\delta_n(\vec{x})| \leq \varepsilon'_n\frac{1}{2}\left(\sum_{k=1}^{r} x_k\right)^2 \leq \frac{r}{2}\varepsilon'_n\sum_{k=1}^{r} x_k^2$$

and

$$0 < \varepsilon'_n \leq C_{11}\frac{\Delta_{s_n}}{s_n\mu_{s_n}} + C_{12}\frac{\sqrt{s_n}\cdot\lambda_{s_n}}{\sqrt{n\mu_{s_n}\Delta_{s_n}}}.$$

For n sufficiently large one could write

$$|\delta_n(\vec{x})| \leq \frac{\gamma}{2}(\vec{x}, \vec{x}), \quad 0 < \gamma < 1. \tag{1.29}$$

Utilizing (1.28) and (1.29) we arrive at

$$|L_n(x)| \le (2\pi)^{-r/2} \exp\left\{-\frac{1}{2}(\vec{x}, \vec{x})\right\}|e^{\delta_n(x)} - 1| \le$$

$$\le |\delta_n(\vec{x})|(2\pi)^{-r/2} \exp\left\{|\delta_n(\vec{x})| - \frac{1}{2}(\vec{x}, \vec{x})\right\} \le$$

$$\le \frac{|\delta_n(\vec{x})|}{(2\pi)^{r/2}} \exp\left\{-\left(\frac{1}{2} - \frac{\gamma}{2}\right)(\vec{x}, \vec{x})\right\},$$

since $|e^x - 1| \le |x|e^{|x|}$ for all x.

Thus we have

$$|\int_{H_n}^{\infty} \exp\{-(\tau_n, \vec{x})\sqrt{n}\}L_n(\vec{x})\, d\vec{x} \le (2\pi)^{-r/2} \int_{H_n}^{\infty} \exp\{-(\tau_n, \vec{x})\sqrt{n}\}| \times$$

$$\times |\delta_n(\vec{x})| \exp\left\{-\left(\frac{1}{2} - \frac{\gamma}{2}\right)(\vec{x}, \vec{x})\right\} d\vec{x} \le (2\pi)^{-r/2} \cdot r \cdot \varepsilon_n' \times$$

$$\times \int_{H_n}^{\infty} \exp\left\{-\lambda_{s_n}(x_1 + \dots + x_r) - \frac{\alpha}{2}(\vec{x}, \vec{x})\right\}(\vec{x}, \vec{x})\, d\vec{x}, \tag{1.30}$$

where $\alpha = 1 - \gamma$.

Since

$$\int_{\lambda_{s_n} - \sqrt{n}\cdot m_i}^{\infty} \exp\left\{-\lambda_{s_n} t - \frac{\alpha}{2}t^2\right\} dt = O(\lambda_{s_n}^{-1}) \tag{1.31}$$

and

$$\int_{\lambda_{s_n} - \sqrt{n}\cdot m_i}^{\infty} t^2 \exp\left\{-\lambda_{s_n} t - \frac{\alpha}{2}t^2\right\} dt = O(\lambda_{s_n})$$

uniformly in i, $1 \le i \le r$, we obtain from (1.30) that

$$\int_{H_n}^{\infty} \exp\left\{-\lambda_{s_n}(x_1 + \dots + x_r) - \frac{\alpha}{2}(\vec{x}, \vec{x})\right\}(\vec{x}, \vec{x})\, d\vec{x} = O(\lambda_{s_n}^{-r+2}). \tag{1.32}$$

Relations (1.22), (1.30) and (1.32) imply that

$$\exp\{n[\log R_n - (\tau_n, \bar{m}_n)]\}|\int_{H_n}^{\infty} \exp\{-(\tau_n, \vec{x})\sqrt{n}\}|L_n(\vec{x})|\, d\vec{x}| \le$$

$$\le \frac{r\varepsilon_n'}{(2\pi)^{r/2}} C_{13} \frac{1}{\lambda_{s_n}^{r-2}} \exp\left\{-\frac{r}{2}\lambda_{s_n}^2\right\} =$$

$$= C_{14}\varepsilon_n'\lambda_{s_n}^2 \left(\frac{\exp\left(-\frac{\lambda_s^2}{2}\right)}{\lambda_{s_n}}\right)^r =$$

$$= C_{14}\varepsilon''_n \left(\frac{1}{\sqrt{2\pi}} \int_{\lambda_{s_n}}^{\infty} \exp\left\{ -\frac{x^2}{2} \right\} dx \right)^r , \tag{1.33}$$

where

$$\varepsilon''_n = C_{15} \frac{\lambda_{s_n}^2 \Delta_{s_n}}{s_n \mu_{s_n}} + C_{16} \frac{\lambda_{s_n}^3 \sqrt{s_n}}{\sqrt{n \mu_{s_n} \Delta_{s_n}}}$$

On the other hand it is easy to verify that

$$I_3 = \exp[n(\log R_n - (\tau_n, \bar{m}_n))](2\pi)^{-r/2} \int_{H_n}^{\infty} \exp\left\{ -\lambda_{s_n} \times \right.$$
$$\left. \times (x_1 + \ldots + x_r) - \frac{1}{2}(\vec{x}, \vec{x}) \right\} d\vec{x} = \exp\left[n(\log R_n - \right.$$
$$\left. - (\tau_n, \bar{m}_n)) + \frac{r}{2}\lambda_{s_n}^2 \right] \left(\frac{1}{\sqrt{2\pi}} \int_{\lambda_{s_n}}^{\infty} \exp\left\{ -\frac{t^2}{2} \right\} dt \right)^r \times$$
$$\times \left[1 + O\left(\frac{\Delta_{s_n} \lambda_{s_n}^2}{s_n \mu_{s_n}} \right) + O\left(\frac{\sqrt{s_n} \cdot \lambda_{s_n}}{\sqrt{n \mu_{s_n} \Delta_{s_n}}} \right) \right]^r . \tag{1.34}$$

Next, taking (1.22) and (1.34) into account we obtain, after obvious simplifications that

$$I_3 = \left(\frac{1}{\sqrt{2\pi}} \int_{\lambda_{s_n}}^{\infty} \exp\left\{ -\frac{t^2}{2} \right\} dt \right)^r \left[1 + O\left(\frac{\Delta_{s_n} \lambda_{s_n}^2}{s_n \mu_{s_n}} \right) + \right.$$
$$\left. + O\left(\frac{\sqrt{s_n} \cdot \lambda_{s_n}^3}{\sqrt{n \mu_{s_n} \Delta_{s_n}}} \right) \right] . \tag{1.35}$$

Thus, in view of (1.23), (1.33) and (1.35) it follows from (1.20) that for n sufficiently large

$$P'_n = \left(\frac{1}{\sqrt{2\pi}} \int_{\lambda_{s_n}}^{\infty} \exp\left\{ -\frac{t^2}{2} \right\} dt \right)^r \left[1 + O\left(\frac{\Delta_{s_n} \lambda_{s_n}^2}{s_n \mu_{s_n}} \right) + \right.$$
$$\left. + O\left(\frac{\sqrt{s_n} \cdot \lambda_{s_n}^m}{\sqrt{n \mu_{s_n} \Delta_{s_n}}} \right) \right]. \tag{1.36}$$

Finally using the same arguments as in N. V. Smirnov's paper [45] we obtain from (1.36) that

$$P_n = P\left\{n^{-1/2}\left|\sum_{k=1}^{n} \xi_n^{(k)}(t_j)\right| > \lambda_{s_n}, \quad j = \overline{1, r}\right\} =$$

$$= \left(\frac{2}{\sqrt{2\pi}} \int_{\lambda_{s_n}}^{\infty} \exp\left\{-\frac{x^2}{2}\right\} dx\right)^r \left[1 + O\left(\frac{\Delta_{s_n}\lambda_{s_n}^2}{s_n \mu_{s_n}}\right) + \right.$$

$$\left. + O\left(\frac{\sqrt{s_n}\cdot\lambda_{s_n}^m}{\sqrt{n\mu_{s_n}\Delta_{s_n}}}\right)\right], \tag{1.37}$$

where $m = \max(3, r)$. The lemma is thus proved.

4. Auxiliary Theorems

Let a sequence of series of events $E_i^{(n)}$, $i = 1, 2, \ldots, s_n$, $n = 1, 2, \ldots$ be given where $s_n \to \infty$ as $n \to \infty$. Denote by $P_{\alpha_1, \alpha_2, \ldots, \alpha_r}$ $(1 \leq \alpha_1 < \alpha_2 < \ldots < \alpha_r < s_n)$ the probability of simultaneous occurrence of the events $E_{\alpha_1}^{(n)}, \ldots, E_{\alpha_r}^{(n)}$ and set

$$\nu_r(s_n) = \sum_{1 \leq \alpha_1 < \alpha_2 < \ldots < \alpha_r \leq s_n} P_{\alpha_1, \alpha_2, \ldots, \alpha_r}.$$

Next, let $P_{s_n}(r)$ be the probability of exactly r positive outcomes among the s_n events and $P_\lambda(r)$ be the Poisson distribution with parameter λ.

LEMMA 1.4 (R. von Mises [46]). If

$$\sum_{t=0}^{s_n} t^k P_{s_n}(t) \to \sum_{t=0}^{\infty} t^k P_\lambda(t),$$

$k = 0, 1, 2, \ldots$, as $n \to \infty$ then

$$\lim_{n\to\infty} P_{s_n}(t) = P_\lambda(t) = \frac{\lambda^t}{t!} e^{-\lambda}. \tag{1.38}$$

The method of proving the next lemma is analogous to the one in the proof of N. V. Smirnov's lemma [47] (Lemma 1, p. 363).

LEMMA 1.5. If for any fixed r, $r = 0, 1, 2, \ldots$, the condition

$$\lim_{n\to\infty} \nu_r(s_n) = \frac{\lambda^r}{r!}$$

is fulfilled then

$$\lim_{n\to\infty} \sum_{t=0}^{s_n} t^k P_{s_n}(t) = \sum_{t=0}^{\infty} t^k P_\lambda(t), \qquad k = 0, 1, 2 \ldots$$

<u>Proof</u>. Introduce the probability generating function

$$\phi_{s_n}(t) = \sum_{r=0}^{s_n} P_{s_n}(r) t^r.$$

Denote by z_k the indicator function of the event $E_k^{(n)}$. Then

$$P_{s_n}(r) = E\left\{ \sum_{i_1<\ldots<i_r} \prod_{k=1}^{r} z_{i_k} \sum_{\ell=1}^{s_n-r} (1 - z_{j_\ell}) \right\},$$

where $i_1, \ldots, i_r$ form a group of r numbers chosen from the sequence $1, \ldots, s_n$ and $j_1, \ldots, j_{s_n-r}$ the complementary group. Since

$$\nu_r(s_n) = E\left\{ \sum_{1\le\alpha_1<\ldots<\alpha_r\le s_n} z_{\alpha_1\ldots\alpha_r} \right\},$$

where $z_{\alpha_1\ldots\alpha_r}$ is the indicator function of the event $\bigcap_1^r E_{\alpha_i}^{(n)}$ it follows that

$$\phi_{s_n}(t) = E\left\{ \prod_{i=1}^{s_n} [z_i t + (1 - z_i)] \right\} = \sum_{r=0}^{s_n} \nu_r(s_n)(t - 1)^r.$$

Hence

$$\sum_{r=0}^{s_n} P_{s_n}(r)(t - 1)^r = \sum_{r=0}^{s_n} \nu_r(s_n) t^r.$$

Comparing the coefficients at the same powers of t in the last equation we obtain

$$\nu_r(s) = \sum_{m=r}^{s_n} C_m^r P_{s_n}(m).$$

In view of the condition of the lemma

$$\frac{\lambda^r}{r!} = \lim_{n\to\infty} \nu_r(s_n) = \lim_{n\to\infty} \sum_{m=r}^{s_n} C_m^r P_{s_n}(m).$$

On the other hand

$$\sum_{m=r}^{\infty} C_m^r P_\lambda(m) = \frac{\lambda^r}{r!}.$$

Consequently

$$\lim_{n\to\infty} \sum_{m=r}^{s_n} C_m^r P_{s_n}(m) = \sum_{m=r}^{\infty} C_m^r P_\lambda(m). \tag{1.39}$$

Since the factorial moments of the r-th order of distributions $\{P_{s_n}(m)\}$ and $\{P_\lambda(m)\}$ are linear functions of the moments of the first r orders of the corresponding distributions, relationship (1.39) implies the assertion of the lemma.

5. Construction of Confidence Regions for Probability Densities

Denote by $E_k^{(n)}$ the event that $|\xi_n(t_k)| \geq \lambda_{s_n}$ ($k = 1, 2, \ldots, s_n$) and set

$$\nu_r(s_n) = \sum_{1\leq\alpha_1<\alpha_2<\ldots<\alpha_r\leq s_n} P(E_{\alpha_1}^{(n)} E_{\alpha_2}^{(n)} \ldots E_{\alpha_r}^{(n)}).$$

Then utilizing the representation of P_n as given in Lemma 1.3 and Remark 1 we obtain

$$\nu_r(s_n) = C_{s_n}^r \left[\left(\frac{2}{\sqrt{2\pi}} \int_{\lambda_{s_n}}^{\infty} \exp\left\{-\frac{t^2}{2}\right\} dt\right)^r \left(1 + O\left(\frac{\Delta_{s_n}\lambda_{s_n}^2}{s_n\mu_{s_n}}\right) + O\left(\frac{\sqrt{s_n}\cdot\lambda_{s_n}^m}{\sqrt{n\mu_{s_n}}\cdot\Delta_{s_n}}\right)\right)\right]. \tag{1.40}$$

LEMMA 1.6. If the conditions of Lemma 1.3 are fulfilled for $\lambda_{s_n} = \ell_{s_n} + \lambda/\ell_{s_n}$ where ℓ_{s_n} is the root of the equation

$$1/s_n = (2\pi)^{-1/2} \int_{\ell_{s_n}}^{\infty} \exp\left\{-\frac{x^2}{2}\right\} dx$$

and $-\infty < \lambda < \infty$, then

$$\nu_r(s_n) \to \nu_r = \frac{1}{r!}(2e^{-\lambda})^r, \quad r = 0, 1, 2, \ldots$$

as n increases.

Proof. Note that

$$\ell_{s_n} = \sqrt{2\log s_n} - \frac{\log\log s_n + \log 4\pi}{2\sqrt{2\log s_n}} + O\left(\frac{1}{\log s_n}\right)$$

(cf. [48] p. 375, eq. (28.6.13)) i.e. $\lambda_{s_n} = O(\sqrt{\log s_n})$ and

$$\frac{1}{\sqrt{2\pi}}\int_{\lambda_{s_n}}^{\infty}\exp\left\{-\frac{t^2}{2}\right\}dt=\frac{\exp\left\{-\frac{\lambda_{s_n}^2}{2}\right\}}{\sqrt{2\pi}\cdot\lambda_{s_n}}(1+O(\lambda_{s_n}^{-2})).$$

To prove the lemma it is sufficient to assert that

$$\lim_{n\to\infty} C_{s_n}^r\left(\frac{2}{\sqrt{2\pi}}\int_{\lambda_{s_n}}^{\infty}\exp\left\{-\frac{x^2}{2}\right\}dx\right)^r=\nu_r, \tag{1.41}$$

to verify (1.41) it is sufficient to establish that

$$\lim_{n\to\infty}\frac{s_n}{\sqrt{2\pi}}\int_{\lambda_{s_n}}^{\infty}\exp\left\{-\frac{x^2}{2}\right\}dx=e^{-\lambda}.$$

Evidently,

$$\frac{s_n}{\sqrt{2\pi}}\int_{\lambda_{s_n}}^{\infty}\exp\left\{-\frac{x^2}{2}\right\}dx=\frac{s_n}{\sqrt{2\pi}\cdot\lambda_{s_n}}\exp\left\{-\frac{\lambda_{s_n}^2}{2}\right\}[1+O(\lambda_{s_n}^{-2})]=$$

$$=e^{-\lambda}\left[1+O\left(\frac{1}{\log s_n}\right)\right].$$

Hence

$$C_{s_n}^r\left(\frac{2}{\sqrt{2\pi}}\int_{\lambda_{s_n}}^{\infty}\exp\left\{-\frac{x^2}{2}\right\}\right)^r=\frac{1}{r!}(2e^{-\lambda})^r\left[1+O\left(\frac{1}{\log s_n}\right)\right]^r.$$

The lemma is thus proved.

The quantity λ_{s_n} is of the order $\sqrt{\log s_n}$; therefore the conditions of the lemma are reduced to

$$\frac{\Delta_{s_n}}{s_n\mu_{s_n}}\log s_n\to 0,\qquad \frac{s_n(\log s_n)^m}{n\mu_{s_n}\Delta_{s_n}}\to 0\quad\text{as}\quad n\to\infty.$$

Define the functions $T_n(x)$ and $\sigma_n(x)$ on the set $R\backslash E_{s_{n+1}}$ in the following manner:

$$T_n(x)=f_n(t_i)\quad\text{and}\quad \sigma_n(x)=(na_n^{-1}Df_n(t_i))^{1/2}\quad\text{for } x\in E_i.$$

Denote by $V_n(\lambda)$ the number of departures of $T_n(x)$ on the set $R\backslash E_{s_{n+1}}$ outside the boundary of the strip bounded by the curves

$$y_n^+(x) = ET_n(x) + \left(\ell_{s_n} + \frac{\lambda}{\ell_{s_n}}\right)\sigma_n(x),$$

$$y_n^-(x) = ET_n(x) - \left(\ell_{s_n} + \frac{\lambda}{\ell_{s_n}}\right)\sigma_n(x).$$

THEOREM 1.1. Under our assumptions 1° and 3° on $f(x)$ and $K(x)$ and assuming that as $n \to \infty$ $(\Delta_{s_n}/s_n\mu_{s_n}) \log s_n \to 0$ and for any fixed m, $m = 0, 1, 2, \ldots,$

$$\frac{s_n(\log s_n)^m}{n\mu_{s_n}\Delta_{s_n}} \to 0 \quad \text{as} \quad n \to \infty \tag{1.42}$$

we have

$$P(V_n(\lambda) = r\} \to \frac{(2e^{-\lambda})^r}{r!} e^{-2e^{-\lambda}}, \quad r = 0, 1, 2, \ldots$$

as n increases.

Proof. It follows from Lemmas 1.3 and 1.6 that $\lim_{n\to\infty} \nu_r(s_n) = (2e^{-\lambda})^r/r!$, $r = 0, 1, 2, \ldots$. The assertion of the theorem now follows directly from Lemma 1.5 and von Mises' Lemma 1.4.

THEOREM 1.2. Under the conditions of Theorem 1.1

$$\lim_{n\to\infty} P\left\{\sqrt{\frac{a}{a_n}} \max_{x \in R\setminus E_{s_n+1}} \left|\frac{T_n(x) - ET_n(x)}{\sigma_n(x)}\right| < \ell_{s_n} + \frac{\lambda}{\ell_{s_n}}\right\} = e^{-2e^{-\lambda}}.$$

The proof follows from Theorem 1.1 since we have on the l.h.s. $P\{V_n(\lambda) = 0\}$.

Remark 2. Let

$$K(x) = \begin{cases} 1/2 & |x| \le 1 \\ 0, & |x| > 1. \end{cases}$$

Then

$$f_n(x) = \frac{1}{2} a_n(F_n(x + a_n^{-1}) - F_n(x - a_n^{-1})),$$

where $F_n(x)$ is the empirical distribution function constructed from the sample. Let the intervals E_i, $i = \overline{1, s_n}$ be of the same length: $h_1 = h_2 = \ldots = h_{s_n} = 2a_n^{-1}$. Then $T_n(x) = f_n(t_k) = f_n^*(x) = (m_k/2n)a_n$, $x \in E_k$ where $f_n^*(x)$ is the usual 'histogram' and m_k, $k = \overline{1, s_n}$ denote the number of observations falling respectively into E_k. From Theorem 1.2 it is easy to deduce a theorem which is a generalization of N. V. Smirnov's theorem 1 in [2]:

THEOREM 1.3. If as n and s_n increase

$$\frac{\log s_n}{a_n \mu_{s_n}} \to 0, \qquad \frac{a_n (\log s_n)^m}{n \mu_{s_n}} \to 0, \qquad m = 0, 1, \ldots,$$

then

$$\lim_{n\to\infty} P\left\{ \sqrt{\frac{n}{a_n}} \max_{x \in R \setminus E_{s_n+1}} \left| \frac{f_n^*(x) - \bar{f}(x)}{(\bar{f}(x))^{1/2}} \right| < \ell_{s_n} + \frac{\lambda}{\ell_{s_n}} \right\} = e^{-2e^{-\lambda}},$$

where

$$\bar{f}(x) = \frac{1}{2a_n} \int_{E_k} f(t)\, dt, \qquad x \in E_k, \qquad k = \overline{1, s_n}.$$

THEOREM 1.4. Under our assumptions 1°, 2° and 4° on f(x) and K(x) and for

$$\frac{\Delta_{s_n}}{s_n \mu_{s_n}} \log s_n \to 0, \qquad \frac{s_n (\log s_n)^m}{n \mu_{s_n} \Delta_{s_n}} \to 0, \qquad \frac{n (\log s_n) \Delta_{s_n}^3}{\mu_{s_n} \cdot s_n^3} \to 0 \qquad (1.43)$$

for any fixed m (m = 0, 1, ...) we have

$$\lim_{n\to\infty} P\left\{ \sqrt{\frac{n}{a_n}} \max_{x \in R \setminus E_{s_n+1}} |T_n(x) - f(x)|\, f^{-1/2}(x) < \right.$$

$$\left. < \sigma \left(\ell_{s_n} + \frac{\lambda}{\ell_{s_n}} \right) \right\} = e^{-2e^{-\lambda}},$$

where

$$\sigma = \left(\int K^2(u)\, du\right)^{1/2}.$$

Proof. Let $b_i < b_{i+1}$ be the end-points of the intervals E_i, $i = \overline{1, s_n}$. Using Taylor's formula we have

$$f(x) = f(b_i) + (x - b_i) f'(\Theta_i), \qquad x \in E_i, \qquad b_i < \Theta_i < b_{i+1},$$

and also

$$ET_n(x) = a_n \int K(a_n(t_i - u)) f(u)\, du = f(b_i) + O\left(\frac{\Delta_{s_n}}{s_n}\right).$$

Consequently,

$$f(x) - ET_n(x) = O\left(\frac{\Delta_{s_n}}{s_n}\right), \quad x \in E_i. \tag{1.44}$$

Analogously we obtain for the function $\sigma(x) = (f(x))^{1/2}\sigma$

$$\sigma_n(x) - \sigma(x) = O\left(\frac{\Delta_{s_n}}{s_n}\right), \quad x \in E_i. \tag{1.45}$$

Next, it can be shown as it was done in Lemma 1.1 that for n sufficiently large, $\sigma_n(x)$ is uniformly bounded from below by a positive quantity:

$$\sigma_n(x) \geq \mu_{s_n}^{1/2}\left(\sigma^2 - M^2\frac{\Delta_{s_n}}{\mu_{s_n}\cdot s_n}\right)^{1/2}. \tag{1.46}$$

Set

$$M_n^* = \max_{x \in R\backslash E_{s_{n+1}}}\left|\frac{T_n(x) - ET_n(x)}{\sigma_n(x)}\right|,$$

$$M_n^{**} = \max_{x \in R\backslash E_{s_{n+1}}}\left|\frac{T_n(x) - f(x)}{\sigma(x)}\right|,$$

$$\eta_n = M_n^{**} - M_n^*$$

and show that

$$P\{|\eta_n| > \varepsilon/\ell_{s_n}\sqrt{na_n^{-1}}\} \to 0 \quad \text{as} \quad n \to \infty.$$

Indeed we have

$$\left|\frac{|T_n(x) - ET_n(x)|}{\sigma_n(x)} - \frac{|T_n(x) - f(x)|}{\sigma(x)}\right| \leq \left|\frac{T_n(x) - ET_n(x)}{\sigma_n(x)} - \right.$$

$$\left. - \frac{T_n(x) - f(x)}{\sigma(x)}\right| \leq \left|\frac{f(x) - ET_n(x)}{\sigma_n(x)}\right| + |T_n(x) - f(x)| \times$$

$$\times \left|\frac{\sigma(x) - \sigma_n(x)}{\sigma_n(x)\sigma(x)}\right|. \tag{1.47}$$

Utilizing Theorem 2.1.10 and the uniform convergence of $f_n(x)$ to $f(x)$ it is easy to verify that $|T_n(x) - f(x)| \to 0$ a.s. uniformly in x. Furthermore, taking (1.44)-(1.46) into account we obtain from (1.47) that $|\eta_n| \leq c_{17}\,\Delta_{s_n}/\sqrt{\mu_n}\cdot s_n$ with probability one. Next we note that in view of the last inequality the probability $\vec{P}\{|\eta_n| \geq C_{17} (\Delta_{s_n} / \sqrt{\mu_{s_n}} \cdot s_n$ is arbitrarily small for large n and s_n. However, for any $\varepsilon > 0$ the difference

$$\frac{\varepsilon\sqrt{a_n}}{\ell_{s_n}\sqrt{n}} - C_{17}\frac{\Delta_{s_n}}{\mu_{s_n}^{1/2} s_n} = \frac{\varepsilon\sqrt{a_n}}{\ell_{s_n}\sqrt{n}}\left(1 - C_{17}\frac{\Delta_{s_n}\sqrt{n}\cdot\ell_{s_n}}{s_n\mu_{s_n}^{1/2}\sqrt{a_n}}\right)$$

in view of condition

$$\frac{n(\log s_n)\Delta_{s_n}^3}{s_n^3\mu_{s_n}} \to 0 \quad \text{as} \quad n\to\infty$$

is positive and therefore

$$P\{|\eta_n| > \varepsilon/\ell_{s_n}\sqrt{na_n^{-1}}\} \to 0 \quad \text{as} \quad n\to\infty.$$

Taking inequalities

$$P\left\{M_n^{**} \leq \frac{\ell_{s_n} + \lambda/\ell_{s_n}}{\sqrt{na_n^{-1}}}\right\} \geq P\left\{M_n^{*} < \frac{\ell_{s_n} + \frac{\lambda - \varepsilon}{\ell_{s_n}}}{\sqrt{na_n^{-1}}}\right\} - $$

$$- P\left\{|\eta_n| > \frac{\varepsilon\sqrt{a_n}}{\ell_{s_n}\sqrt{n}}\right\},$$

$$P\left\{M_n^{**} < \frac{\ell_{s_n} + \frac{\lambda}{\ell_{s_n}}}{\sqrt{na_n^{-1}}}\right\} \leq P\left\{M_n^{*} < \frac{\ell_{s_n} + \frac{\lambda + \varepsilon}{\ell_{s_n}}}{\sqrt{na_n^{-1}}}\right\} +$$

$$+ \vec{P}\left\{|\eta_n| > \frac{\varepsilon\sqrt{a_n}}{\ell_{s_n}\sqrt{n}}\right\},$$

as well as Theorem 1.2 into account we obtain

$$e^{-2e^{-(\lambda-\varepsilon)}} \leq \varliminf_{n\to\infty} P\left\{M_n^{**} \leq \frac{\ell_{s_n} + \frac{\lambda}{\ell_{s_n}}}{\sqrt{na_n^{-1}}}\right\} \leq$$

$$\overline{\lim_{n\to\infty}} P\left\{M_n^{**} \leq \frac{\ell_{s_n} + \frac{\lambda}{\ell_{s_n}}}{\sqrt{na_n^{-1}}}\right\} \leq e^{-2e^{-(\lambda+\varepsilon)}}.$$

Whence, since ε is arbitrarily small, it follows that

$$\lim_{n\to\infty} P\left\{\sqrt{\frac{n}{a_n}}\, M_n^{**} \leq \ell_{s_n} + \frac{\lambda}{\ell_{s_n}}\right\} = e^{-2e^{-\lambda}}.$$

The theorem is thus proved.

With the aid of Theorem 1.4 one can easily solve the problem of constructing a confidence region for the theoretical probability density corresponding to a given confidence coefficient $\alpha\,(0 < \alpha < 1)$: given α, one first determines $\lambda = \lambda_\alpha$ satisfying the equation $\exp\{-\,2\exp-\lambda\} = \alpha$. Set

$$\tau_\alpha = \left(\ell_{s_n} + \frac{\lambda_\alpha}{\ell_{s_n}}\right)\frac{\sigma}{\sqrt{na_n^{-1}}}.$$

Then the inequality

$$\max_{x\in R\setminus E_{s_{n+1}}} \left|\frac{T_n(x) - f(x)}{\sqrt{f(x)}}\right| \leq \tau_\alpha$$

implies that with probability as close as desirable to α the region $G_\alpha(K)$ bounded by the curves

$$S_n^+(x) = T_n(x) + \tau_\alpha^2/2 + \tau_\alpha(T_n(x) + \tau_\alpha^2/4)^{1/2},$$

$$S_n^-(x) = T_n(x) + \tau_\alpha^2/2 - \tau_\alpha(T_n(x) + \tau_\alpha^2/4)^{1/2},$$

covers on the set $R\setminus E_{s_{n+1}}$ the unknown density $f(x)$.

6. Examples Clarifying the Meaning of Conditions in Theorems 1.1 and 1.4

EXAMPLE 1. Let $f(x)$ be the normal density

$$f(x) = \frac{1}{\sqrt{2\pi}}\, e^{-x^2/2}.$$

Let $E_{s_{n+1}}$ be the set complementary to the interval $[-\,b_{s_n}, b_{s_n}]$. In this case $\Delta_{s_n} = 2b_{s_n}$ and $\mu_{s_n} = (2\pi)^{-1/2}\exp\{-\,(b_{s_n}^2/2)\}$. If we set for example $b_{s_n} = (\log s_n)^{1/2}$ and $s_n = n^\beta$ then the conditions (1.42) will be consistent for $0 < \beta < 2/3$ and conditions (1.43) for $2/5 < \beta < 2/3$.

EXAMPLE 2. Let

$$f(x) = A(p, q)x^{p-1}(1 - x)^{q-1} \quad \text{and} \quad R\backslash E_{s_{n+1}} = [\eta_{s_n}, 1 - \eta_{s_n}],$$

$$\eta_{s_n} \to 0 .$$

In this case Δ_{s_n} is bounded and $\mu_{s_n} = O(\eta_{s_n}^{\gamma})$, $\gamma = \max(p - 1, q - 1)$, $1 < p, q < 2$. Let $\eta_{s_n} = s_n^{-1/2}$, $s_n = [n^{\beta}]$. Conditions (1.42) are consistent if $0 < \beta < (2 + \gamma)^{-1}$ and conditions (1.43) for $2(6 - \gamma)^{-1} < \beta < 2(2 + \gamma)^{-1}$ and $2 \leq p, q < 3$.

7. Approximating a Probability Density in the Regular Case

Assume that we are dealing with approximating a density on a finite interval $[a, b]$ which does not depend on n, and let $[a, b] = R\backslash E_{s_{n+1}} = \bigcup_{k=1}^{s_n} E_k$ and $\min_{a\leq x\leq b} f(x) \geq \mu_0 > 0$ (the regular case). Then the conditions (1.43) of Theorem 1.4 are simplified: the first relation $\Delta_{s_n}(s_n\mu_n)^{-1} \cdot \log s_n \to 0$ drops out and the others become $n^{-1}s_n(\log s_n)^m \to 0$, $s_n^{-3}n \log s_n \to 0$.

Assume that $K(x)$ is a bounded probability density which satisfies 5°:

$$K(x) = K(- x), \quad K(|x_2|) \leq K(|x_1|) \quad \text{for} \quad |x_2| > |x_1| ,$$

$$\int u^2K(u)\, du < \infty.$$

Let $\min_{a\leq x\leq b} f(x) \geq \mu_0 > 0$ and the interval $[a, b]$ be subdivided into s_n subintervals E_i, $i = \overline{1, s}$ of equal length $h_n = (b - a)/s_n$.

LEMMA 1.7.

$$\varepsilon_n = \max_{1\leq i,j\leq s_n, i\neq j} |E(\xi_n(t_i)\xi_n(t_j))| \leq C_{18}K^*(h_na_n) + O(a_n^{-1}), \tag{1.48}$$

where

$$K^*(x) = \int K(u)K(x - u)\, du.$$

Proof. As $n \to \infty$, $Ef_n(x) \to f(x)$ and

$$a_n \int K^2(a_n(x - u))f(u)\, du \to f(x) \int K^2(u)\, du$$

uniformly in x, $a \leq x \leq b$ (cf. [5]). Therefore for any $\varepsilon > 0$ there exists an $N(\varepsilon)$ such that

$$a_n \int K^2(a_n(t_i - u)f(u)\, du - a_n^{-1}(Ef_n(t_i))^2 \geq \mu_0 \int K^2(u)\, du - \varepsilon$$

$$\text{as} \quad n > N(\varepsilon).$$

Whence in view of condition 5° we have

$$\varepsilon_n \leq \frac{1}{(\mu_0 \int K^2(u)\, du - \varepsilon)} \sup_x f(x)[\int K(u)K(a_n|t_i - t_j| - u)\, du +$$

$$+ O(a_n^{-1})] \leq C_{18}K^*(h_n a_n) + O(a_n^{-1}),$$

as since $\min\limits_{1\leq i,j\leq s_n, i\neq j} |t_i - t_j| \geq h_n$. The lemma is thus proved.

The following lemma follows directly from the regularity condition.

LEMMA 1.8.

$$\max_{1\leq i\leq n} \sup_{a\leq x\leq b} |\xi_n^{(i)}(x)| \leq C_{19} a_n^{1/2} \quad \text{a.s.}$$

The following analog of Lemma 1.3 is valid.

LEMMA 1.9. Let λ_{s_n} be non-negative and an increasing function of s_n. If as $n \to \infty$

$$\varepsilon_n \lambda_{s_n}^2 \to 0, \qquad \left(\frac{a_n}{n}\right)^{1/2} \lambda_{s_n}^m \to 0,$$

then

$$P\{|\xi_n(t_i)| \geq \lambda_{s_n}, \quad i = \overline{1, r}\} = \left[\frac{2}{\sqrt{2\pi}} \int_{\lambda_{s_n}}^{\infty} \exp\left\{-\frac{x^2}{2}\right\} dx\right]^r \times$$

$$1 + O(\varepsilon_n \lambda_{s_n}^2) + O\left(\left(\frac{a_n}{n}\right)^{1/2} \lambda_{s_n}^m\right),$$

where $m = \max(3, r)$.

The proof is completely analogous to the proof of Lemma 1.3 and is based on Lemma 1.8. From Lemma 1.9 one can deduce the following analogs of Theorems 1.2 and 1.4, respectively.

THEOREM 1.5. Under assumptions 1° and 5° on the functions $f(x)$ and $K(x)$ and moreover as $n \to \infty$

$$\varepsilon_n \log s_n \to 0 \quad \text{and} \quad \sqrt{\frac{a_n}{n}}(\log s_n)^m \to 0 \quad \text{for any fixed } m,$$

$$m = 0, 1, 2, \ldots .$$

We have

$$\lim_{n\to\infty} P\left\{\sqrt{\frac{n}{a_n}} \max_{a\leq x\leq b} \left|\frac{T_n^{(x)} - ET_n(x)}{\sigma_n(x)}\right| < \ell_{s_n} + \frac{\lambda}{\ell_{s_n}}\right\} = e^{-2e^{-\lambda}}.$$

THEOREM 1.6. Under assumptions 1°, 2° and 5° on the functions $f(x)$ and $K(x)$ and moreover as $n \to \infty$

$$\varepsilon_n \log s_n \to 0, \quad \left(\frac{a_n}{n}\right)^{1/2} (\log s_n)^m \to 0, \text{ and } \quad \frac{n \log s_n}{s_n^2 a_n} \to 0,$$

for any fixed m, $m = 0, 1, 2, \ldots$. We have

$$\lim_{n\to\infty} P\left\{\sqrt{\frac{a}{a_n}} \max_{a \le x \le b} \left| \frac{T_n(x) - f(x)}{(f(x))^{1/2}} \right| < \sigma\left(\ell_{s_n} + \frac{\lambda}{\ell_{s_n}}\right)\right\} = e^{-2e^{-\lambda}}.$$

We shall now investigate the problem of refining the condition $(n \log s_n)/s_n^2 a_n) \to 0$. (In the case when $T_n(x)$ is used as an approximating function for $f(x)$ one cannot improve on this condition even if $f(x)$ is of higher order of smoothness than the one stipulated in Theorem 1.6). For this purpose we shall define a continuous function $\tilde{f}_n(x)$ on the interval $[a + h_n/2, b - h_n/2]$

$$\tilde{f}_n(x) = f_n(t_k) + \frac{x - t_k}{h_n} [f_n(t_{k+1}) - f_n(t_k)], \quad t_k \le x \le t_{k+1},$$

$$k = \overline{1, s_n - 1}, \tag{1.49}$$

where t_k is the center of the interval E_k.

Since $t_k = b_k - (h_n/2)$ and $t_{k+1} = b_k + (h_n/2)$, $k = \overline{1, s_n - 1}$, where $b_k < b_{k+1}$ are the endpoints of the intervals Δ_k, we obtain from (1.49) that

$$\tilde{f}_n(x) = \frac{1}{2} [f_n(b_k - h_n/2) + f_n(b_k + h_n/2)] + \frac{x - b_k}{h_n} \times$$

$$\times [f_n(b_k + h_n/2) - f_n(b_k - h_n/2)], \quad t_k \le x \le t_{k+1}.$$

Analogously we introduce the function $q_n(x)$ by setting

$$q_n(x) = \frac{1}{2} [\phi_n(b_k - h_n/2) + \phi_n(b_k + h_n/2)] + \frac{x - b_k}{h_n} \times$$

$$\times [\phi_n(b_k + h_n/2) - \phi(b_k - h_n/2)], \quad t_k \le x \le t_{k+1}$$

where

$$\phi_n(y) = (a_n \int K^2((y - u)a_n) f(u)\, du)^{1/2}.$$

Assume that $f(x)$ satisfies condition 3°. Since $K(x) = K(-x)$ one can easily verify the following relations

$$Ef_n(x) - f(x) = O(h_n^2) = O(s_n^{-2}), \quad x \in [t_k, t_{k+1}], \tag{1.50}$$

$$(f(x) \int K^2(x)\, dx)^{1/2} - q_n(x) = O(s_n^{-2}), \quad x \in [t_k, t_{k+1}]. \quad (1.51)$$

Consider the quantity

$$M_n = \max_{a+h_n/2 \leq x \leq b-h_n/2} \left| \frac{\tilde{f}_n(x) - E\tilde{f}_n(x)}{q_n(x)} \right| .$$

We shall now show that the function $|(\tilde{f}_n(x) - Ef_n(x))q_n^{-1}(x)|$ attains its maximum at one of the end points of the interval $[t_k, t_{k+1}]$, $k = \overline{1, s_n - 1}$. Indeed let

$$z = \frac{x - t_k}{h_n}, \quad 0 \leq z \leq 1,$$

then

$$(\tilde{f}_n(x) - Ef_n(x))q_n^{-1}(x) = [(1 - z)A + zB][(1 - z)C + zD]^{-1} = \Psi(z),$$

where

$$A = f_n(t_k) - Ef_n(t_k), \quad B = f_n(t_{k+1}) - Ef_n(t_{k+1}), \quad C = q(t_k),$$

$$D = q(t_{k+1})$$

Whence

$$\Psi'(z) = \frac{BC - AD}{[C + z(D - C)]^2} .$$

This means that $\Psi(z)$, $0 \leq z \leq 1$ is either increasing or decreasing depending on the sign of $BC - AD$. Thus

$$M_n = \max_{1 \leq k \leq s_n} \left| \frac{f_n(t_k) - Ef_n(t_k)}{\phi_n(t_k)} \right| .$$

In this case Theorem 1.5 implies

THEOREM 1.7. Let $f(x)$ possess a bounded derivative of the second order on $[a, b]$ and $K(x)$ satisfies condition 5°. If, with the increase in n,

$$\varepsilon_n \log s_n \to 0, \quad a_n(\log s_n)^m n^{-1} \to 0, \quad ns_n^{-4}a_n^{-1} \log s_n \to 0,$$

then

$$\lim_{n\to\infty} P\left\{ \sqrt{\frac{n}{a_n}} \max_{x\in[a+h_n/2, b-h_n/2]} \left| \frac{\tilde{f}_n(x) - f(x)}{f^{1/2}(x)} \right| < \right.$$

$$\left. < \sigma(\ell_{s_n} + \lambda/\ell_{s_n}) \right\} = e^{-2e^{-\lambda}},$$

where $\sigma^2 = \int K^2(u)\, du$.

Proof. Denote

$M_n^* = \max_{a\le x\le b} |(\tilde f_n(x) - f(x))f^{-1/2}(x)|$ and let $\eta_n = M_n^* - M_n$. From a chain of inequalities analogous to the one presented in the proof of Theorem 1.4 and in view of (1.50) and (1.51) we have $|\eta_n| \le C_{20}s_n^{-2}$. However, by the condition of the theorem the difference

$$\varepsilon_n \sqrt{a_n}\, \ell_{s_n}^{-1} n^{-1/2} - C_{20} s_n^{-2}$$

is positive for n large.

Hence

$$P\left\{ |\eta_n| > \frac{\varepsilon\sqrt{a_n}}{\sqrt{n}\cdot \ell_{s_n}} \right\} \to 0 \quad \text{as} \quad n \to \infty.$$

From now on, the proof of the theorem is completely analogous to the proof of Theorem 1.4 and is therefore omitted.

In the same manner as in Theorem 1.4 one can determine from here a confidence region with a given confidence coefficient α for the unknown density. This region is determined by the same curves $S_n^+(x)$ and $S_n^-(x)$ where, however, $T_n(x)$ is replaced by $\tilde f_n(x)$. It follows from the expressions for $S_n^+(x)$ and $S_n^-(x)$ that to determine the optimal form of the kernel $K(x)$ which minimizes the confidence region for $f(x)$ it is sufficient to obtain the minimum of the functional $\int K^2(x)\,dx$ for fixed a_n, α and the additional condition $\int u^2K(u)\,du = 1$. This problem belongs to isoparametric problems of a constrained calculus of variation. Euler's equation for variational problems of this kind is written in the form $K(y) - \lambda_1 + \lambda_2 y^2 = 0$. Parameters λ_1 and λ_2 are determined from the conditions imposed on $K(y)$. We thus obtain the optimal form of the kernel $K(y)$ (see M. G. Kendall and A. Stuart, 'The Advanced Theory of Statistics, Vol 2, Inference and Relationship', p. 666)

$$K_0(x) = \begin{cases} \dfrac{3}{4\sqrt{5}} - \dfrac{3x^2}{20\sqrt{5}} & \text{for } |x| \le \sqrt{5}, \\ 0 & \text{for } |x| \ge \sqrt{5}, \end{cases} \tag{1.52}$$

Thus the minimal confidence region G_α^0 which covers the unknown density $f(x)$ on the whole interval $(a + h_n/2, b - h_n/2)$ with probability as close to α as desired is bounded by the curves

$$S_n^+(x) = \tilde f_n(x) + \frac{\tau_\alpha^{0\,2}}{2} + \tau_\alpha^0(\tilde f_n(x) + \tau_\alpha^{0\,2}/4)^{1/2},$$

$$\bar{S}_n^0(x) = \tilde{f}_n(x) + \tau_\alpha^{0\,2}/2 - \tau_\alpha^0(\tilde{f}_n(x) + \tau_\alpha^{0\,2}/4)^{1/2},$$

where

$$\tau_\alpha^0 = (\ell_{s_n} + \lambda_\alpha/\ell_{s_n}) \frac{\sqrt{3.5^{-3/2}}}{\sqrt{na_n^{-1}}}.$$

8. Consider the problem as to when the conditions of Theorems 1.5-1.7 are fulfilled. In view of (1.48) we have for ε_n the bound

$$\varepsilon_n \leq C_{21}K^*(a_n h_n) + C_{22}a_n^{-1}$$

where

$$K^*(u) = \int K(x)K(u - x)\, dx.$$

It follows from property 5° that

$$\lim_{|x| \to \infty} |x| K^*(x) = 0.$$

Indeed,

$$\begin{aligned} |x| K^*(x) &\leq \int |x - u| K(u)K(x - u)\, du + \int |u| K(u)K(x - u)\, du \leq \\ &\leq 2 \int |u| K(u)K(x - u)\, du \leq 2 \sup_{|u| \leq \delta} K(x - u) \times \\ &\times \int |u| K(u)\, du + 2 \sup_{u \in R} K(u) \int_{|u| > \delta} |u| K(u)\, du. \end{aligned}$$

Selecting first δ sufficiently large and then x large enough we obtain that for an $\varepsilon > 0$ $|x| K^*(x) \leq \varepsilon$ for $|x| > A(\varepsilon)$. Then

$$\varepsilon_n \leq C_{21}K^*(a_n h_n) + C_{22} \frac{1}{a_n h_n} h_n = 0\left(\frac{1}{a_n h_n}\right).$$

Set $s_n = An^\alpha$, $a_n = Bn^\beta$. The conditions of Theorem 1.5 are fulfilled if

$$0 < \alpha < \beta < 1.$$

The conditions of Theorem 1.6 are fulfilled if

$$0 < \alpha < \beta < 1, \quad 2\alpha + \beta > 1.$$

The conditions of Theorem 1.7 are fulfilled if

$$0 < \alpha < \beta < 1, \quad 4\alpha + \beta > 1.$$

If $K(x) \equiv 0$ for $|x| \geq A_0$, $0 < A_0 < \infty$ and $h_n = 2A_0 a_n^{-1}$ then the conditions of Theorems 1.5-1.7 are simplified in the sense that the condition $\varepsilon_n \log s_n \to 0$ is now unnecessary since $\varepsilon_n = O(s_n^{-1})$.

Remark 3. The problem of the limiting distribution of the maximal deviation of $f_n(x)$ from $f(x)$ in the regular case was later considered in paper [49]. Under quite stringent conditions on $f(x)$, $K(x)$ and a_n the limiting distribution of this statistics was obtained. A complete expansion of the distribution was derived in [50]; in the maximal deviation $f(x)$ is also replaced by a step function.

2. Limiting Distribution of Quadratic Deviation of Two Non-Parametric Kernel-Type Density Estimators

1. Let a sample $X_1, X_2, \ldots, X_n$ of n independent observations of a random variable X with an unknown density $f(x)$ be given. Consider a nonparametric estimator $f_n(x)$ of the density $f(x)$ of the type (1.1.1):

$$f_n(x) = n^{-1}a_n \sum_{i=1}^{n} K(a_n(x - X_i)).$$

Function $K(x)$ satisfies the following regularity conditions: 1°. It vanishes outside a finite interval $(-A, A)$ and together with its derivative is continuous on this interval; or 2°. It is absolutely continuous on $R = (-\infty, \infty)$ $x^2K(x) \in L_1$ and $K'(x) \in L_1$. Moreover, in both cases $\int K(x)\,dx = 1$ and $K(x) = K(-x)$.

The set of these functions will be denoted by G and we shall denote by Π the set of densities possessing continuous and bounded derivatives up to the second order. Let $F_n(x)$ be the empirical sample distribution function and $F(x)$ be the distribution function determined by the density $f(x)$.

Introduce the empirical process

$$\xi_n(t) = \sqrt{n}\,(S_n(t) - t), \quad 0 \le t \le 1,$$

where $S_n(t) = F_n(F^{-1}(t))$ and the Brownian bridge

$$W^0(t) = W(t) - tW(1), \quad 0 \le t \le 1,$$

where $W(t)$ is the standard Wiener process on [0,1]. Our subsequent arguments are based on the following theorem.

THEOREM 2. [51]. There exists a probability space (Ω, F, P) on which versions of $S_n(t)$ and a sequence of Brownian bridges $W_n^0(t)$ can be defined such that

$$P(\sup_{0\le t\le 1} |n(S_n(t) - t) - n^{1/2}W_n^0(t)| > C \log n + x) \le C_1 e^{-\lambda x}, \tag{2.1}$$

for all $x > 0$ where C_1, λ and C are positive constants.

The following asymptotic expansion for $F_n(x)$ can be easily derived from (2.1):

$$F_n(x) = F(x) + \frac{1}{\sqrt{n}} W_n^0(F(x)) + \frac{\log n}{n} Z_n(x) \qquad (2.2)$$

where $Z_n(x)$ is a random function such that

$$P\{\sup_x |Z_n(x)| > C + \tau\} \le C_1 n^{-\lambda\tau}.$$

From here for a suitable choice of $\tau > 0$ it follows that $\sup_x |Z_n(x)|$ is bounded a.s. The expansion presented in (2.2) is of interest on its own. It can be used, for example, for obtaining several well known asymptotic properties of the empirical distribution function. Moreover, it allows us to obtain an expansion of the normalized deviation

$$\left(\frac{n}{a_n}\right)^{1/2}(f_n(x) - Ef_n(x)): (n/a_n)^{1/2}(f_n(x) - Ef_n(x)) =$$
$$= a_n^{1/2} \int K(a_n(x-u))\, dW_n^0(F(u)) + e_n(x), \qquad (2.3)$$

where

$$e_n(x) = \log n/\sqrt{na_n^{-1}} \int K(a_n(x-u))\, dZ_n(u).$$

Since $Z_n(x)$ is uniformly bounded in x and $K'(x) \in L_1$ it follows from (2.3) that

$$\left(\frac{n}{a_n}\right)^{1/2} f_n(x) - Ef_n(x)) = a_n^{1/2} \int K(a_n(x-u))\, dW_n^0(F(u)) +$$
$$+ O_p\left(\frac{\log n}{\sqrt{na_n^{-1}}}\right). \qquad (2.4)$$

(Recall that a random variable ξ_n is $O_p(b_n)$ if for an $\varepsilon > 0$ there exists a $C(\varepsilon)$ and a number $n(\varepsilon)$ such that $P\{|\xi_n| < b_n C(\varepsilon)\} > 1 - \varepsilon$ for $n > n(\varepsilon)$). Utilizing (2.4) one can easily obtain the assertion that the random vector

$$\left(\left(\frac{n}{a_n}\right)^{1/2}(f_n(x_j) - Ef_n(x_j)), \quad j = \overline{1, r}\right),$$

where x_j, $j = \overline{1, r}$ are fixed points in $R = (-\infty, \infty)$ and $x_j \ne x_i$, $i \ne j$, possesses asymptotically multivariate normal distribution (O, C) with a zero vector of the mean values and the covariance matrix $C = (C_{ij})$,

$$C_{ii} = f(x_i) \int K^2(u)\, du, \quad C_{ij} = 0, \quad i \ne j.$$

In paper [49] the limiting distribution of quadratic deviation

$\int (f_n(x) - f(x))^2 r(x)\,dx$ was investigated. Here Brillinger's approximation [52] of the empirical process by means of a sequence of Brownian bridges

$$\sup_{0\le t\le 1} |n^{1/2}(S_n(t)^{-t} - W_n^0(t)| = O_p(n^{-1/4}(\log n)^{1/2}(\log\log n)^{1/4}).$$

was utilized. This estimate is evidently quite weak as compared to (2.2). In this connection in paper [49] a rather strong condition

$$n^{-1/4}(\log n)^{1/2}(\log\log n)^{1/4} = O(a_n^{-1})$$

was imposed on the sequence $\{a_n\}$.

The aim of this section is to obtain the limiting distribution of the quadratic deviation of two nonparametric kernel-type estimators of a density constructed from two independent samples.

2. Let two independent samples X_{ij} $(j = \overline{1, n_i},\ i = 1, 2)$ from populations with densities $f_1(x)$ and $f_2(x)$ respectively be given. Based on these samples we construct two non parametric estimators

$$f_{n_i}(x) = \frac{a_i}{n_i} \sum_{j=1}^{n_i} K(a_i(x - X_{ij})), \quad i = 1, 2.$$

To compare these estimators $f_{n_1}(x)$ and $f_{n_2}(x)$ we introduce the quantity

$$U_{n_1 n_2} = \frac{N_1 N_2}{N_1 + N_2} \int (f_{n_1}(x) - f_{n_2}(x))^2 r(x)\,dx, \quad N_i = \frac{n_i}{a_i},$$
$$i = 1, 2,$$

where the function $r(x)$ is piecewise continuous, bounded and $r(x) \in L_1(R)$. Assume that the hypothesis H_0 is valid according to which $f_1(x) = f_2(x) = f_0(x)$ ($f_0(x)$ is a given density); the sample sizes n_1 and n_2 increase indefinitely in such a manner that

$$\frac{n_2}{n_1} = \tau, \quad a_1 = a_2 = a_n, \quad n = \frac{n_2}{1 + \tau}$$

Denote

$$U^{(1)}_{n_1 n_2} = \frac{N_1 N_2}{N_1 + N_2} \int (\tilde f_{n_1}(x) - \tilde f_{n_2}(x))^2 r(x)\,dx,$$

where

$$\tilde f_{n_i}(x) = f_{n_i}(x) - Ef_{n_i}(x), \quad i = 1, 2.$$

LEMMA 2.1. Let $f_0(x) \in \Pi$ and $K(x) \in G$. If $na_n^{-4} \to 0$ as $n \to \infty$ the following relation is valid:

$$U_{n_1 n_2} - U_{n_1 n_2}^{(1)} = O_p(a_n^{-1}). \tag{2.5}$$

Proof. We have

$$U_{n_1 n_2} - U_{n_1 n_2}^{(1)} = \frac{N_1 N_2}{N_1 + N_2} \int (Ef_{n_1}(x) - Ef_{n_2}(x))^2 r(x)\, dx +$$

$$+ 2\frac{N_1 N_2}{N_1 + N_2} \int \tilde{f}_{n_1}(x)(Ef_{n_1}(x) - Ef_{n_2}(x)) r(x)\, dx +$$

$$+ 2\frac{N_1 N_2}{N_1 + N_2} \int \tilde{f}_{n_2}(x)(Ef_{n_1}(x) - Ef_{n_2}(x)) r(x)\, dx =$$

$$= A_1 + A_2 + A_3.$$

Since

$$Ef_{n_i}(x) = f_0(x) + O(a_n^{-2}), \quad i = 1, 2,$$

it follows that

$$A_1 = O(na_n^{-5}) = O(a_n^{-1}).$$

We now estimate $E|A_2|$. In view of

$$\operatorname{cov}(\tilde{f}_{n_1}(x), \tilde{f}_{n_1}(y)) = n_1^{-1} a_n^2 \{ \int K(a_n(x-u)) K(a_n(y-u)) f_0(u) \times$$

$$\times du - \int K(a_n(x-u)) f_0(u)\, du \int K(a_n(y-u)) f_0(u)\, du \},$$

we have

$$E|A_2| \le 2\frac{N_1 N_2}{N_1 + N_2} \{ \int | \int K(a_n(x-u))[Ef_{n_1}(x) - Ef_{n_2}(x)] \times$$

$$\times r(x)\, dx |^2 f_0(u)\, du\, n_1^{-1} a_n^2 \}^{1/2} \le C_2 \sqrt{n} \cdot a_n^{-3} = O(a_n^{-1}),$$

Analogously it can be shown that

$$E|A_3| = O(a_n^{-1}).$$

Hence

$$U_{n_1 n_2} - U_{n_1 n_2}^{(1)} = O_p(a_n^{-1}).$$

The lemma is thus proved.

Set

$$\xi_{n_i}(x) = a_n^{1/2} \int K(a_n(x-u))\, dW^0_{n_i}(F_0(x)), \quad i = 1, 2,$$

where $W^0_{n_i}$ and $W^0_{n_2}$ are independent sequences of Brownian bridges described in Theorem 2.1 based on sample sizes n_1 and n_2 respectively. Then in view of (2.4)

$$\left(\frac{N_1 N_2}{N_1 + N_2}\right)^{1/2} (\tilde{f}_{n_1}(x) - \tilde{f}_{n_2}(x)) = \alpha_2 \xi_{n_1}(x) - \alpha_1 \xi_{n_2}(x) + $$

$$+ O_p\left(\frac{\log n}{\sqrt{n a_n^{-1}}}\right). \tag{2.6}$$

where

$$\alpha_i = \left(\frac{N_i}{N_1 + N_2}\right)^{1/2}, \quad i = 1, 2.$$

We estimate the variance of the quantity

$$\int (\alpha_2 \xi_{n_1}(x) - \alpha_1 \xi_{n_2}(x)) r(x)\, dx.$$

It follows from the independence of $W^0_{n_1}$ and $W^0_{n_2}$ that

$$D \int (\alpha_2 \xi_{n_1}(x) - \alpha_1 \xi_{n_2}(x)) r(x)\, dx = \alpha_2^2 \iint E\xi_{n_1}(x)\xi_{n_1}(y) r(x) \times$$

$$\times r(y)\, dx\, dy + \alpha_1^2 \iint E\xi_{n_2}(x)\xi_{n_2}(y) r(x) r(y)\, dx\, dy.$$

It is easy to verify that

$$E\xi_{n_i}(x)\xi_{n_j}(y) = a_n[\int K(a_n(x-u))K(a_n(y-u)) f_0(u)\, du - \int K \times$$

$$\times (a_n(x-u)) f_0(u)\, du \int K(a_n(y-u)) f_0(u)\, du] .$$

Whence after some simple manipulations we obtain

$$\iint E\xi_{n_i}(x)\xi_{n_i}(y) r(x) r(y)\, dx\, dy = O(a_n^{-1}). \tag{2.7}$$

Hence

$$D \int (\alpha_2 \xi_{n_1}(x) - \alpha_1 \xi_{n_2}(x)) r(x)\, dx = O(a_n^{-1}). \tag{2.8}$$

Representation (2.6) and relation (2.8) yield

$$U^{(1)}_{n_1n_2} = \int (\alpha_2\xi_{n_1}(x) - \alpha_1\xi_{n_2}(x))^2 r(x)\, dx + O_p\left(\frac{\log^2 n}{na_n^{-1}}\right). \quad (2.9)$$

Denote

$$\eta_{n_i}(x) = a_n^{1/2} \int K(a_n(x-u))\, dW_{n_i}(F_0(u)),$$

$$U^{(2)}_{n_1n_2} = \int (\alpha_2\xi_{n_1}(x) - \alpha_1\xi_{n_2}(x))^2 r(x)\, dx,$$

$$U^{(3)}_{n_1n_2} = \int (\alpha_2\eta_{n_1}(x) - \alpha_1\eta_{n_2}(x))^2 r(x)\, dx,$$

$$\varepsilon_{n_i}(x) = a_n^{1/2} \int K(a_n(x-u))dF_0(u)\, W_{n_i}(1).$$

Then

$$U^{(2)}_{n_1n_2} - U^{(3)}_{n_1n_2} = O_p(a_n^{-1}). \quad (2.10)$$

Indeed,

$$E|U^{(2)}_{n_1n_2} - U^{(3)}_{n_1n_2}| \le 2E\left|\int (\alpha_2\eta_{n_1}(x) - \alpha_1\eta_{n_2}(x))(\alpha_1\varepsilon_{n_2}(x) - \right.$$

$$\left. - \alpha_2\varepsilon_{n_1}(x))r(x)\, dx\right| + E\int (\alpha_2\varepsilon_{n_1}(x) - \alpha_1\varepsilon_{n_2}(x))^2 r(x)\, dx =$$

$$= A_1 + A_2. \quad (2.11)$$

It is easy to verify that

$$A_2 = \alpha_2^2\, E\int \varepsilon^2_{n_1}(x)r(x)\, dx + \alpha_1^2\, E\int (\varepsilon_{n_2}(x))^2 r(x)\, dx =$$

$$= a_n E(\alpha_2^2\, W^2_{n_1}(1) + \alpha_1^2\, W^2_{n_2}(1)) \int \left[\int K(a_n(x-u))f_0(u)\, du\right]^2 \times$$

$$\times\, r(x)\, dx \le C_3 a_n^{-1}. \quad (2.12)$$

We now proceed to estimate A_1. We have

$$A_1 \le 2E\left|\int \alpha_2\eta_{n_1}(x)(\alpha_1\varepsilon_{n_2}(x) - \alpha_2\varepsilon_{n_1}(x))r(x)\, dx\right| +$$

$$+ 2E\left|\int \alpha_1\eta_{n_2}(x)(\alpha_1\varepsilon_{n_2}(x) - \alpha_2\varepsilon_{n_1}(x))r(x)\, dx\right| = I_1 + I_2.$$

We bound I_1.

$$I_1 = a_n^{-1}\, E\left|(\alpha_1 W_{n_2}(1) - \alpha_2 W_{n_1}(1)) \int \{\alpha_2 \int \phi(u - a_n^{-1}v) r \times\right.$$

$$\left. \times\, (u - a_n^{-1}v)K(v)\, dv\}\, dW_{n_1}(F_0(u))\right|.$$

However,

$$E\left(\int (\alpha_2 \int \phi(u - a_n^{-1}v) r(u - a_n^{-1}v) K(v)\, dv)\, dW_{n_1}(F_0(u))\right)^2 \le C_4,$$

where

$$\phi(t) = \int K(u) f_0(t - ua_n^{-1})\, du,$$

whence

$$I_1 = O(a_n^{-1}). \tag{2.13}$$

Analogously,

$$I_2 = O(a_n^{-1}). \tag{2.14}$$

Taking (2.13) and (2.14) into account we obtain

$$A_2 = O(a_n^{-1}). \tag{2.15}$$

Hence in view of (2.12), (2.15) and (2.11) we have

$$E\left|U^{(2)}_{n_1n_2} - U^{(3)}_{n_1n_2}\right| = O(a_n^{-1}). \tag{2.16}$$

The bound (2.16) yields (2.10).

Finally combining (2.5), (2.9) and (2.10) we arrive at

$$U_{n_1n_2} - U^{(3)}_{n_1n_2} = O_p(a_n^{-1}) + O_p\left(\frac{\log^2 n}{na_n^{-1}}\right). \tag{2.17}$$

We now proceed to investigate the limiting distribution of the functional

$$U^{(3)}_{n_1n_2} = \int [\alpha_2\eta_{n_1}(x) - \alpha_1\eta_{n_2}(x)]^2 r(x)\, dx.$$

We find the covariance functions $R_n(x, y)$ of the Gaussian process $\alpha_2\eta_{n_1}(x) - \alpha_1\eta_{n_2}(x)$. Since the processes $\eta_{n_1}(x)$ and $\eta_{n_2}(x)$ are independent we have

$$R_n(x, y) = \alpha_2^2\, E\eta_{n_1}(x)\eta_{n_1}(y) + \alpha_1^2\, E\eta_{n_2}(x)\eta_{n_2}(y).$$

However,

$$\begin{aligned} E\eta_{n_i}(x)\eta_{n_i}(y) &= \int K(z) K(a_n(x - y) + z) f_0(x - a_n^{-1}z)\, dz = \\ &= f_0(x) \int K(z) K(a_n(x - y) + z)\, dz + O(a_n^{-1}) = \\ &= f_0(x) K_0(a_n(x - y)) + O(a_n^{-1}) \end{aligned}$$

where the bound $O(\cdot)$ is uniform in x, y and $K_0 = K*K$.

Hence

$$R_n(x, y) = f_0(x) K_0(a_n(x - y)) + O(a_n^{-1}). \tag{2.18}$$

The cumulant $\chi_{n_1n_2}(m)$ of order m of the random variable $U^{(3)}_{n_1n_2}$ is given by a formula presented in paper [49] (It also follows from the formula given in Gikhman and Skorokhod's book: The Theory of Stochastic Processes I [53] Chapter V, Section 6, p. 351.)

$$\chi_{n_1n_2}(m) = (m-1)!\,2^{m-1} \int \ldots \int \prod_{i=1}^{m} R_n(u_i, u_{i+1}) r(u_i)\, du_i,$$

$$u_{m+1} = u_1. \tag{2.19}$$

We have

$$EU^{(3)}_{n_1n_2} = \chi^{(1)}_{n_1n_2} = \int f_0(x) r(x)\, dx \int K^2(x)\, dx + O(a_n^{-1}) \tag{2.20}$$

and

$$DU^{(3)}_{n_1n_2} = \chi_{n_1n_2}(2) = 2a_n^{-1} \int f_0^2(x) r^2(x)\, dx \int K_0^2(x)\, dx + O(a_n^{-2}).$$

Introduce the transformation

$$u_s = t_m + a_n^{-1} \sum_{i=s}^{m-1} t_i, \quad s = \overline{1, m-1} \tag{2.21}$$

and $u_m = t_m$. The Jacobian of the transformation equals $(a_n^{-1})^{m-1}$. Applying this transformation to (2.19) one can show that the m-th cumulant $\chi_{n_1n_2}(m)$ with precision up to the terms of higher order of smallness is equal to

$$(m-1)!\,2^{m-1}(a_n^{-1})^{m-1}[K{*}K]^{(m)}(0) \int f_0^m(x) r^m(x)\, dx, \tag{2.22}$$

where $[K{*}K]^m(u)$ denotes the m-fold convolution of $K_0(x)$ with itself.

THEOREM 2.2. Under our assumptions on $f_0(x)$ and $K(x)$ and as $na_n^{-2} \to \infty$, $na_n^{-4} \to 0$ the random variable

$$a_n^{1/2}(U_{n_1n_2} - \int f_0(x) r(x)\, dx \int K^2(x)\, dx)$$

is asymptotically normal with mean 0 and variance

$$\sigma^2 = 2 \int f_0^2(x) r^2(x)\, dx \int K_0^2(x)\, dx.$$

Proof. It follows from relations (2.20) and (2.22) that

$$a_n^{1/2}(U^{(3)}_{n_1n_2} - \int f_0(x) r(x)\, dx \int K^2(x)\, dx)$$

is distributed in the limit normally with mean 0 and variance σ^2. On the other hand since

$$a_n^{3/2} \log^2 n/n = (na_n^{-2})^{-1}(na_n^{-4})^{1/8} n^{-1/8} \log^2 n \to 0$$

as $n \to \infty$ it follows from (2.17) that

$$a_n^{1/2}(U_{n_1n_2} - U_{n_1n_2}^{(3)}) = o_p(1).$$

Hence

$$a_n^{1/2}(U_{n_1n_2} - \int f_0(x)r(x)\,dx \int K^2(x)\,dx)$$

is distributed normal $(0, \sigma)$ which completes the proof.

Theorem 2.2 allows us to construct an asymptotic set at level α for testing the hypothesis H_0 that $f_1(x) = f_2(x) = f_0(x)$ (where $f_0(x)$ is a given density belonging to the class Π). To do this it is only required to compute $U_{n_1n_2}$ and reject H_0 if

$$U_{n_1n_2} \geq d_n(\alpha) = \int f_0(x)r(x)\,dx \int K^2(x)\,dx + a_n^{-1/2}\lambda_\alpha\sigma,$$

where λ_α is the quantile at level α of the standard normal distribution.

We note that statistic $U_{n_1n_2}$ is normalized by means of quantities which depend on the probability density. If, however the hypothesis does not determine $f_0(x)$ one should then replace in the normalizing constant the quantities

$$\int f_0(x)r(x)\,dx, \quad \int f_0^2(x)r^2(x)\,dx$$

by

$$\int f_{n_1n_2}(x)r(x)\,dx, \quad \int f_{n_1n_2}^2(x)r^2(x)\,dx, \quad f_{n_1n_2}(x) =$$

$$= (n_1 + n_2)^{-1}[n_1 f_{n_1}(x) + n_2 f_{n_2}(x)].$$

Denote

$$A_1 = a_n^{1/2} \int |f_{n_1n_2}(x) - f_0(x)|\,r(x)\,dx,$$

$$A_2 = \int |f_{n_1n_2}^2(x) - f_0^2(x)|\,r^2(x)\,dx.$$

We have

$$EA_1 \leq a_n^{1/2} E \int |f_{n_1n_2}(x) - Ef_{n_1n_2}(x)|\,r(x)\,dx + O(a_n^{-3/2}) \leq$$

$$\leq C_1 a_n^{1/2}\{\int \vec{E}(f_{n_1n_2}(x) - \vec{E}f_{n_1n_2}(x))^2 r(x)\,dx\}^{1/2} +$$

$$+ O(a_n^{-3/2}) = O\left(\frac{a_n}{\sqrt{n}}\right).$$

Hence

$$A_1 = o_p(1).$$

$$A_2 \leq C_5 \sup_{-\infty<x<\infty} |f_{n_1n_2}(x) - Ef_{n_1n_2}(x)| + O(a_n^{-2}).$$

However, in view of (2.1.5) $\sup_{-\infty<x<\infty} |f_{n_1n_2}(x) - Ef_{n_1n_2}(x)| = O_p(1)$. Thus $A_2 = o_p(1)$.

Summarizing that stated above we obtain the following corollary to Theorem 2.2.

COROLLARY. The random variable

$$\sigma_n^{-1} a_n^{1/2} (U_{n_1n_2} - \int f_{n_1n_2}(x) r(x)\, dx \int K^2(x)\, dx)$$

is asymptotically normal with mean 0 and variance 1. Here

$$\sigma_n^2 = 2 \int f_{n_1n_2}^2(x) r^2(x)\, dx \int K_0^2(x)\, dx.$$

With the aid of this corollary one can construct a test for testing the hypothesis H_0: $f_1(x) = f_2(x)$; the critical region for testing this hypothesis is determined by the inequality $U_{n_1n_2} \geq \tilde{d}_n(\alpha)$ where

$$\tilde{d}_n(\alpha) = \int f_{n_1n_2}(x) r(x)\, dx \int K^2(x)\, dx + a_n^{-1/2} \lambda_\alpha \sigma_n.$$

2. Assume now that the null hypothesis H_0 is false and that the hypothesis

$$H_1: f_1(x) = f_0(x) + \gamma_n \phi_1(x), \quad f_2(x) = f_0(x) + \gamma_n \phi_2(x)$$

$$\int \phi_i(x)\, dx = 0, \quad i = 1, 2, \quad \gamma_n \downarrow 0 \quad \text{as} \quad n \to \infty \tag{2.23}$$

is valid.

THEOREM 2.3. Let $K(x) \in G$, $f_i(x) \in \Pi$, $i = 1, 2$. If $a_n = n^\delta$ and $\gamma_n = n^{-(1/2)+\delta/4}$, $1/4 < \delta < 1/2$ then the statistic

$$a_n^{1/2} (U_{n_1n_2} - \int f_0(x) r(x)\, dx \int K^2(x)\, dx)$$

under the alternative H_1 is distributed normally in the limit with parameters

$$(\int [\phi_1(x) - \phi_2(x)]^2 r(x)\, dx, \quad 2 \int f_0^2(x) r^2(x)\, dx \int K_0^2(x)\, dx).$$

Proof. Represent $U_{n_1n_2}$ as the sum

$$U_{n_1n_2} = \frac{N_1N_2}{N_1 + N_2} \int [\tilde{f}_{n_1}(x) - \tilde{f}_{n_2}(x)]^2 r(x)\, dx + 2na_n^{-1} \int [\tilde{f}_{n_1}(x) -$$

$$- \tilde{f}_{n_2}(x)][E_1 f_{n_1}(x) - E_2 f_{n_2}(x)\, r(x)\, dx + na_n^{-1} \int [E_1 f_{n_1}(x) -$$

$$- E_2 f_{n_2}(x)]^2 r(x)\, dx = A_1 + A_2 + A_3, \tag{2.24}$$

where $\tilde{f}_{n_i}(x) = f_{n_i}(x) - E_i f_{n_i}(x)$. Here $E_i f_{n_i}(x)$ denotes the mathematical expectation with respect to $f_i(x)$, $i = 1, 2$.

Tracing down the proof of Theorem 2.2 it is easy to verify that $a_n^{1/2}(A_1 - \int f_0(x)r(x)\,dx \int K^2(x)\,dx)$ is distributed asymptotically normal $(0, \sigma)$.

In the same manner as in the proof of Lemma 2.1 we obtain

$$E|A_2| \leq c_6 \sqrt{n}\, \gamma_n a_n^{-1} = O(a_n^{-1/2})$$

and

$$n a_n^{1/2} A_3 \to \int [\phi_1(x) - \phi_2(x)]^2 r(x)\,dx. \tag{2.25}$$

Thus relations (2.24) and (2.25) imply the assertion of the theorem.

Theorem 2.3 yields

COROLLARY. Under the conditions of Theorem 2.3 the local behavior of the power $P_{H_1}(U_{n_1n_2} \geq \tilde{d}_n(\alpha))$ is as follows:

$$P_{H_1}(U_{n_1n_2} \geq \tilde{d}_n(\alpha)) \to 1 - \Phi(\lambda_\alpha - \frac{1}{\sigma} \int (\phi_1(x) - \phi_2(x))^2 r(x)\,dx.$$

The proof follows from the relations

$$a_n^{1/2}\sigma_n^{-1}(U_{n_1n_2} - \int f_{n_1n_2}(x)r(x)\,dx \int K^2(x)\,dx) = a_n^{1/2}\sigma^{-1}\sigma_n^{-1} \times$$
$$\times\, \sigma(U_{n_1n_2} - \int f_0(x)r(x)\,dx \int K^2(x)\,dx) + a_n^{1/2}\sigma_n^{-1} \times$$
$$\times \int (f_{n_1n_2}(x) - f_0(x))r(x)\,dx \int K^2(x)\,dx$$

and

$$\sigma_n - \sigma = o_p(1), \quad a_n^{1/2} \int (f_{n_1n_2}(x) - f_0(x))r(x)\,dx \times$$
$$\times \int K^2(x)\,dx = o_p(1).$$

It should be noted that with positive weight functions $r(x)$ the test for testing the hypothesis H_0 against the alternative of the form (2.23) is asymptotically strictly unbiased since the mathematical expectation of $\int (\phi_1(x) - \phi_2(x))^2 r(x)\,dx > 0$ and equals zero if and only if $\phi_1(x) = \phi_2(x)$.

We present a simple example of the comparison of the powers of the $U_{n_1n_2}$-test and Pearson's $\chi^2_{n_1n_2}$-test. Let the distributions be concentrated on the interval [0, 1] and be uniform (hypothesis H_0).

Consider a sequence of alternatives:

$$H_1 : f_1(x) = 1 + \gamma_n \phi_1(x), \quad f_2(x) = 1 + \gamma_n \phi_2(x), \quad \gamma_n = n^{-1/2+\delta/4}$$

Let

$$r(x) = 1, \quad x \in [0, 1] \quad \text{and} \quad r(x) = 0, \quad x \notin [0, 1] \quad \text{and}$$

$$K(x) = \begin{cases} 1 & |x| \le 1/2, \\ 0 & |x| > 1/2. \end{cases}$$

Since

$$\sigma^2 = 2 \int K_0^2(x)\, dx = 2 \int_{|x| \le 1} (1 - |x|)^2\, dx = 4/3,$$

it follows from Theorem 2.3 and its corollary that

$$P_{H_1}(U_{n_1 n_2} \ge d_n(\alpha)) \to 1 - \phi\left(\lambda_\alpha - \frac{\sqrt{3}}{2} \int_0^1 (\phi_1(x) - \phi_2(x))^2\, dx \right),$$
$$P_{H_1}(U_{n_1 n_2} \ge \tilde{d}_n(\alpha)) \to 1 - \phi\left(\lambda_\alpha - \frac{\sqrt{3}}{2} \int_0^1 (\phi_1(x) - \phi_2(x))^2\, dx \right) \tag{2.26}$$

as $n \to \infty$.

Subdivide the interval [0, 1] into intervals $\Delta_1, \Delta_2, \ldots, \Delta_{s_n}$; the lengths of the intervals Δ_j are the same: $\Delta_j = \Delta = 1/s_n$, where $s_n = [a_n = n^\delta]$, $1/4 < \delta < 1/2$. Consider Pearson's $\chi^2_{n_1 n_2}$-test:

$$\chi^2_{n_1 n_2} = n_1 n_2 \sum_{j=1}^{s_n} \frac{1}{\mu_j + \nu_j} \left(\frac{\mu_j}{n_1} - \frac{\nu_j}{n_2} \right)^2,$$

where μ_i and ν_j are respectively the number of observations in the first and the second samples belonging to the interval Δ_j.

Utilizing the method of the proof developed in [54] one can show that

$$P_{H_1}(\chi^2_{n_1 n_2} \ge d_n(\alpha) = s_n + \lambda_\alpha \sqrt{2 s_n}) \to$$
$$\to 1 - \phi\left(\lambda_\alpha - \frac{1}{\sqrt{2}} \int (\phi_2(x) - \phi_1(x))^2\, dx \right). \tag{2.27}$$

Comparing (2.26) and (2.27) we conclude that the test-statistic $U_{n_1 n_2}$ is asymptotically more powerful than the test-statistic $\chi^2_{n_1 n_2}$ relative to the alternative hypothesis H_1.

We present yet another comparison. Let $Y_1 \le Y_2 \le \ldots \le Y_{n_2}$, $Y_0 = 0$, $Y_{n_2+1} = 1$ be the observed increasing ordered statistics; let s_i be the number of $X_i^{(1)}$ - s falling into the interval $[Y_{i-1}, Y_i]$. In paper [18] a test for testing the hypothesis H_0 based on the statistic

$$V^2_{n_1n_2} = \sum_{i=1}^{n_2+1} s_i^2, \quad n_2/n_1 = \tau > 0.$$

was investigated. It was proven that the power of this test under the alternative $H_1: g_n(x) = 1 + n^{-1/4}\phi(x)$, $n = n_2/1 + \tau$ approaches a nondegenerate limit, i.e. the power approaches a number which exceeds the level of significance. It is thus evident that for the test statistic $V^2_{n_1n_2}$ alternatives of the form $g_n(x) = 1 + n^{-(1/2-\delta/4)}\phi(x)$ do not differ in the limit from the null hypothesis since $1/2 - \delta/4 > 1/4$, $1/4 < \delta < 1/2$. Hence the test based on the statistic $U_{n_1n_2}$ is more powerful in the limit for the alternative $g_n(x) = 1 + n^{-(1/2 - \delta/4)}\phi(x)$ than the test based on $V^2_{n_1n_2}$.

3. The Asymptotic Power of the $U_{n_1n_2}$-Test in the Case of 'Singular' Close Alternatives

The aim of this section is to compare tests based on estimators of distribution functions and densities.

As it is well known for the tests constructed in terms of empirical distribution functions, the local 'close' alternatives differ from the null hypothesis by a quantity of order $n^{-1/2}$. Therefore it is quite evident that tests based on $U_{n_1n_2}$ for alternatives similar to (3.2.23) are less powerful than the tests of, say, the Kolmogorov-Smirnov-Cramér-von Mises type.

We shall consider new types of sequences of locally close alternatives:

$$H_1: f_1(x) = f_0(x), \quad f_2(x) = f_0(x) + \alpha_n\phi_n(x - \ell),$$

where $\alpha_n \to 0$, $\phi_n(x)$ is twice differentiable and moreover satisfies the conditions:

$$\beta_n = \int \phi_n^2(x)\, dx \to 0, \quad \int \phi_n(x)\, dx = 0,$$

$$\int |\phi_n(x)|\, dx = O(\beta_n), \quad \int_{|x|>\lambda} \phi_n^2(x)\, dx = 0(\beta_n), \quad \lambda > 0,$$

and

$$\beta_n^3 \int (\phi_n''(x))^2\, dx < C_1,$$

where ℓ is a fixed point of continuity of $r(x)$ such that $r(\ell) \neq 0$. Alternatives of this type will be called '<u>singular</u>'.

THEOREM 3.1. Let the hypothesis H_1 be valid, $na_n^{-2} \to \infty$, $na_n^{-4} \to 0$,

$$na_n^{-1/2}\alpha_n^2\beta_n \to C_0 \neq 0, \quad a_n^{-1} = O(\beta_n)$$

and $\alpha_n \cdot \beta_n = O(n^{-1/2})$ as $n \to \infty$; then

a) $$P_{H_1}(U_{n_1n_2} \geq d_n(\alpha)) \to 1 - \phi\left(\lambda_\alpha - \frac{C_0}{\sigma} r(\ell)\right),$$

b) $$P_{H_1}(U_{n_1n_2} \geq \tilde{d}_n(\alpha)) \to 1 - \phi\left(\lambda_\alpha - \frac{C_0}{\sigma} r(\ell)\right),$$

where

$$\sigma^2 = 2 \int f_0^2(x)r^2(x)\, dx \int K_0^2(x)\, dx.$$

Proof. We have

$$U_{n_1n_2} = \frac{N_1N_2}{N_1 + N_2} \int [\tilde{f}_{n_1}(x) - \tilde{f}_{n_2}(x)]^2 r(x)\, dx + 2\frac{N_1N_2}{N_1 + N_2} \times$$

$$\times \int [\tilde{f}_{n_1}(x) - \tilde{f}_{n_2}(x)][E_1f_{n_1}(x) - E_2f_{n_2}(x)]r(x)\, dx +$$

$$+ \frac{N_1N_2}{N_1 + N_2} \int [E_1f_{n_1}(x) - E_2f_{n_2}(x)]^2 r(x)\, dx =$$

$$= A_n^{(1)} + A_n^{(2)} + A_n^{(3)}, \tag{3.1}$$

where $\tilde{f}_{n_i}(x)$ and $E_if_{n_i}$ are defined as in Section 2.

The following inequality results from the condition of the theorem

$$a_n^{1/2}\alpha_n \leq C_2(na_n^{-2})^{-1/2}. \tag{3.2}$$

Taking into account inequality (3.2) and the properties of functionals $U^{(1)}$, $U^{(2)}$ and $U^{(3)}$ of the preceding section, we obtain that under the hypothesis H_1 the random variable

$$(A_n^{(1)} - \int f_0(x)r(x)\, dx \int K^2(x)\, dx)a_n^{1/2}$$

is distributed asymptotically normally $(0, \sigma)$.

We shall estimate $E|A_n^{(2)}|$.

Since

$$\text{cov}\,(\tilde{f}_{n_i}(x),\ \tilde{f}_{n_i}(y)) = n^{-1}a_n^{-2}(\int K(a_n(x - u))K(a_n(y - u))$$

$$f_i(u)\, du - \int K(a_n(x - u))f_i(u)\, du \int K(a_n(y - u))f_i(u)\, du),$$

then

$$E|A_n^{(2)}| \le C_3\sqrt{n}\, a_n^{-1}\alpha_n \{ \int [a_n^{-1} \int K_0(a_n(u - v))\phi_n(v - \ell)\, dv]^2 \times$$

$$\times (f_1(u) + f_2(u))\, du\}^{1/2} \le C_4\sqrt{n}\, a_n^{-1}\alpha_n\beta_n^{1/2}, \tag{3.3}$$

where K = K*K.

However $\sqrt{n}\, a_n^{-1/2}\alpha_n\beta_n^{1/2} \le C_5 a_n^{-1/4}$ for n sufficiently large. Therefore we obtain from (3.3) that

$$E|A_n^{(2)}| = O(a_n^{-1/2}). \tag{3.4}$$

Consider now $A_n^{(3)}$. We have

$$a_n^{1/2}A_n^{(3)} = na_n^{-1/2}\alpha_n^2 \int [\int K(t)\phi_n(x - \ell + ta_n^{-1})\, dt]^2 r(x)\, dx =$$

$$= na_n^{-1/2}\alpha_n^2 \int [\phi_n(x - \ell) + a_n^{-2} \int\int_0^1 t^2K(t)(1 - z) \times$$

$$\times \phi_n''(x - \ell + tza_n^{-1})\, dt\, dz]^2 r(x)\, dx.$$

Simple manipulations show that under our assumptions:

$$na_n^{-1/2}\alpha_n^2 a_n^{-2} \int\int\int_0^1 \phi_n(x - \ell)t^2K(t)(1 - z)\phi_n''(x - \ell + tza_n^{-1}) \times$$

$$\times r(x)\, dt\, dz\, dx = O\left(\left(\frac{a_n^{-1}}{\beta_n}\right)^2\right),$$

$$na_n^{-1/2}\alpha_n^2 a_n^{-4} \int [\int\int_0^1 t^2K(t)(1 - z)\phi_n''(x - \ell + tza_n^{-1}) \times$$

$$\times dt\, dz]^2 r(x)\, dx = O\left(\left(\frac{a_n^{-1}}{\beta_n}\right)^4\right).$$

Therefore we have

$$a_n^{1/2}A_n^{(3)} = na_n^{-1/2}\alpha_n^2 \int \phi_n^2(x - \ell)r(x)\, dx + O\left(\left(\frac{a_n^{-1}}{\beta_n}\right)^2\right). \tag{3.5}$$

Now we shall determine the asymptotic form of $\beta_n^{-1} \int \phi_n^2(x - \ell)r(x)\, dx$.

Let $\varepsilon > 0$. Since r(x) is continuous at point ℓ, there exists $\eta > 0$ such that $|r(x) - r(\ell)| < \varepsilon$ provided $|x - \ell| < \eta$. Therefore

$$|\beta_n^{-1} \int \phi_n^2(x - \ell)[r(x) - r(\ell)]\, dx| \le \beta_n^{-1} \int_{|x-\ell|<\eta} \phi_n^2(x - \ell) \times$$

$$\times |r(x) - r(\ell)|\, dx + \frac{2M}{\beta_n} \int_{|x-\ell|\ge\eta} \phi_n^2(x - \ell)\, dx =$$

$$= \sup_{|x-\ell|<\eta} |r(x) - r(\ell)| + \frac{2M}{\beta_n} \int_{|x|\geq\eta} \phi_n^2(x)\, dx \leq$$

$$\leq \varepsilon + \beta_n^{-1} 2M \int_{|x|\geq\eta} \phi_n^2(x)\, dx = \varepsilon + O(1),$$

where $M = \sup_x r(x)$.

Since ε is arbitrary it follows that

$$\beta_n^{-1} \int \phi_n^2(x - \ell) r(x)\, dx \to r(\ell).$$

Consequently in view of (3.5) we have

$$a_n^{1/2} A_n^{(3)} \to C_0 r(\ell) \quad \text{as} \quad n \to \infty \tag{3.6}$$

From (3.1), (3.4) and (3.6) taking the asymptotic normality of $A_n^{(1)}$ into account we arrive at the assertion of case a) of the theorem. Case b) is proved analogously utilizing case a).

Integrating $f_1(x)$ and $f_2(x)$ we can verify that the alternatives differ from the null hypothesis by a quantity of the order $\alpha_n \cdot \beta_n = o(n^{-\frac{1}{2}})$. This shows that test-statistics based on deviations between sample distribution functions, such as for example, tests of the Kolmogorov-Smirnov-Cramér-von Mises type are unable to distinguish 'singular' hypotheses from the null hypothesis. Therefore tests based on $U_{n_1 n_2}$ under singular alternatives are more powerful than the tests of the Kolmogorov-Smirnov-Cramér-von Mises type.

The function

$$\phi_n(x) = W\left(\frac{x}{\gamma_n}\right) \tag{3.7}$$

may serve as an example of $\phi_n(x)$; here $\gamma_n \to 0$, the function $W(x)$ is bounded and possesses continuous derivatives up to the second order, and, moreover

$$\int W(x)\, dx = 0$$

while $W''(x) \in L_2(-\infty, \infty)$ [56]. Functions $W(x/\gamma_n)$ are of the form of 'peaks' of size $W(0)$ which deflate as n increases. In our case $\beta_n = \gamma_n \int W^2(x)\, dx$. Set $\alpha_n = a_n^{\alpha} n^{-1/2}$, $\gamma_n = a_n^{-\beta}$, where $\beta > \alpha$, $1/2 + \beta = 2\alpha$, $0 < \beta < 1$. Then $a_n^{-1} = O(\beta_n)$, $\alpha_n \cdot \beta_n = O(n^{-1/2})$ and

$$n a_n^{-1/2} \alpha_n^2 \beta_n \to C_0 = \int W^2(x)\, dx.$$

Thus from this particular example (3.7) of function $\phi_n(x)$ we arrive at

COROLLARY

$$P_{H_1}\{U_{n_1 n_2} \geq d_n(\alpha)\} \to 1 - \phi\left(\lambda_\alpha - \frac{r(\ell)}{\sigma}\int W^2(x)\,dx\right),$$

$$P_{H_1}\{U_{n_1 n_2} \geq \tilde{d}_n(\alpha)\} \to 1 - \phi\left(\lambda_\alpha - \frac{r(\ell)}{\sigma}\int W^2(x)\,dx\right).$$

The conditions imposed on α and β are fulfilled if, for example, we set

$$\alpha = \frac{5}{8}, \quad \beta = \frac{3}{4}, \quad \alpha = \frac{2}{3}, \quad \beta = \frac{5}{6}; \quad \alpha = \frac{5}{12}, \quad \beta = \frac{2}{3}$$

and so on.

Remark. It is easy to observe that one could consider an alternative which is a finite sum of deviations $a_n\phi_n(x - \ell_j)$, $j = 1, 2, \ldots$, centered around distinct points ℓ_j, $j = 1, 2, \ldots$, at which $r(x)$ is continuous and $r(\ell_j) \neq 0$.

4. Testing for Symmetry of a Distribution

1. Let a sample $X_1, \ldots, X_n$ from a population X be given whose distribution is defined by the probability density $f(x)$. Assume that $f(x)$ is symmetric with respect to the point $x = 0$, i.e. $f(x) = f(-x)$ for all $x \in (-\infty, \infty)$ (hypothesis H_0). To test this hypothesis we shall construct in this section a test whose statistic is based on a non-parametric estimator of probability density of the form (1.1.1).

Assume that the kernel $K(x) \in G$ and the density $f \in \Pi$ (cf. Section 2 of Chapter 3).

To compare $f_n(x)$ and $f_n(-x)$ we introduce the statistic

$$T_n = na_n^{-1}\frac{1}{2}\int [f_n(x) - f_n(-x)]^2 r(x)\,dx, \tag{4.1}$$

where the function $r(x)$ is piecewise continuous, bounded and integrable and, moreover $r(x) = r(-x)$ for all $x \in (-\infty, \infty)$.

It is easy to verify that $E((f_n(x) - f_n(-x))/H_0) = 0$. Therefore (4.1) can be written as

$$T_n = na_n^{-1}\frac{1}{2}\int (\tilde{f}_n(x) - \tilde{f}_n(-x))^2 r(x)\,dx,$$

where

$$\tilde{f}_n(x) = f_n(x) - Ef_n(x).$$

Denote

$$\xi_n(x) = a_n^{1/2}\int K(a_n(x-u))\,dW_n^0(F(u)),$$

where $\{W_n^0\}$ is a sequence of Brownian bridges (cf. Section 2 of Chapter 3). Then in view of (3.2.4)

$$\sqrt{n}\, a_n^{-1/2}(\tilde{f}_n(x) - \tilde{f}_n(-x)) = \xi_n(x) - \xi_n(-x) + O_p\left(\frac{\log n}{\sqrt{na_n^{-1}}}\right) =$$

$$= \eta_n(x) + O_p\left(\frac{\log n}{\sqrt{na_n^{-1}}}\right), \tag{4.2}$$

where

$$\eta_n(x) = a_n^{1/2} \frac{1}{\sqrt{2}} \int [K(a_n(x - u)) - K(a_n(x + u))]\, dW_n^0(F(u)).$$

The variance of the variable $\int \eta_n(x) r(x)\, dx$ is a linear combination of terms of the form

$$a_n \int\int EK(a_n(x - X))K(a_n(y - X))r(x)r(y)\, dx\, dy = O(a_n^{-1}).$$

This together with (4.2) implies that

$$T_n = \int \eta_n^2(x) r(x)\, dx + O_p(n^{-1} a_n \log^2 n). \tag{4.3}$$

Denote

$$\psi_n(x, u) = K(a_n(x - u)) - K(a_n(x + u)),$$

$$Z_n(x) = \frac{1}{\sqrt{2}} a_n^{1/2} \int \psi_n(x, u)\, dW_n(F(u)),$$

$$T_n^{(1)} = \int \eta_n^2(x) r(x)\, dx, \quad T_n^{(2)} = \int Z_n^2(x) r(x)\, dx,$$

where $W_n(t)$ is the standard Wiener process on [0, 1].

We shall verify that

$$T_n^{(1)} - T_n^{(2)} = o_p(a_n^{-1}). \tag{4.4}$$

Indeed, taking into account that $W_n^0(t) = W_n(t) - tW_n(1)$ we obtain that

$$|T_n^{(1)} - T_n^{(2)}| \le a_n W_n^2(1)\left(\int \left(\int \psi_n(x, u) f(u)\, du\right)^2 r(x)\, dx\right) +$$

$$+ 2a_n |W_n(1)| \left| \int \left(\int\int \psi_n(x, u)\psi_n(x, v) f(u)\, du\, dW_n \times\right.\right.$$

$$\left.\left. \times (F(v))\right) r(x)\, dx \right| = A_1 + A_2,$$

and moreover

$$\int\int \psi_n(x, u)\psi_n(x, v) f(u)\, dW_n(F(v))\, du$$

is a linear combination of terms of the form

$$\int\int K(a_n(x + \alpha_1 u))K(a_n(x + \alpha_2 v)) f(u)\, dW_n(F(v))\, du,$$

where α_1 and α_2 are quantities taking on two values ± 1.

It is easy to verify that

$$EA_1 = O(a_n^{-1}). \tag{4.5}$$

We shall now estimate $\vec{E}A_2$. For this purpose it is sufficient to estimate

$$M_n = E\left| \int \left(\int \int K(a_n(x + \alpha_1 u))K(a_n(x + \alpha_2 v))f(u)\ dW_n \times \right.\right.$$
$$\left.\left. \times\ (F(v))\ du\right) r(x)\ dx \right|.$$

Noting that

$$E\left(\int \left(\int K(y)r(v - ya_n^{-1})\phi(v - ya_n^{-1})\ dy\right)\ dW_n(F(v))\right)^2 \leq C_1,$$

where

$$\phi(t) = \int K(u)f(t - ua_n^{-1})\ du,$$

we arrive at

$$M_n \leq C_2 a_n^{-2},$$

and hence

$$EA_2 = O(a_n^{-1}). \tag{4.6}$$

Thus in view of (4.5) and (4.6) we have (4.4). Finally, combining (4.3) and (4.4) we obtain

$$T_n = \int Z_n^2(x)r(x)\ dx + O_p(a_n^{-1}) + O_p\left(\frac{\log^2 n}{na_n^{-1}} \right). \tag{4.7}$$

2. Asymptotic distribution of the functional

$$T_n^{(2)} = \int Z_n^2(x)r(x)\ dx.$$

We shall obtain the covariance function $R_n(x, y)$ of the Gaussian process $Z_n(x)$. $R_n(x, y)$ is a linear combination of summands of the form $EK(a_n(x + \alpha_1 X))K(a_n(y + \alpha_2 X))$, i.e.

$$R_n(x, y) = a_n \frac{1}{2} \sum_{(\alpha_1, \alpha_2)} \alpha_1\alpha_2 \int K(a_n(x + \alpha_1 u))K(a_n(y + \alpha_2 u)) \times$$
$$\times\ f(u)\ du, \tag{4.8}$$

where the summation over α_1, α_2 is taken over all possible combinations of these symbols taking on two possible values $\pm$ 1. We shall calculate as an example the summand in (4.8) for $\alpha_1 = -1$, $\alpha_2 = +1$:

$$a_n \int K(a_n(x - u))K(a_n(y + u))f(u)\ du =$$
$$= f(x)K_0(a_n(x + y)) + O(a_n^{-1}),$$

where $K_0 = K{*}K$. Other summands are calculated analogously. As the result we obtain

$$R_n(x, y) = \frac{1}{2}[f(x) + f(y)]L_n(x, y) + O(a_n^{-1}), \tag{4.9}$$

where

$$L_n(x, y) = K_0(a_n(x - y)) - K_0(a_n(x + y)).$$

Furthermore, the m-th order cumulant $\chi_n(m)$ of the random variable $T_n^{(2)}$ is given by a formula which is analogous to (3.2.19). We shall now evaluate $ET_n^{(2)}$ and $DT_n^{(2)}$.

We have

$$ET_n^{(2)} = \chi_n(1) = \frac{1}{2}\int R_n(x, x)r(x)\,dx =$$

$$= \int f(x)r(x)\,dx \int K^2(x)\,dx + O(a_n^{-1}), \tag{4.10}$$

$$DT_n^{(2)} = \chi_n(2) = \frac{1}{4}\int\int R_n(x_1, x_2)R_n(x_2, x_1)r(x_1(r(x_2)\,dx_1\,dx_2 =$$

$$= \frac{1}{4}\int\int [f(x_1) + f(x_2)]^2L_n^2(x_1, x_2)r(x_1)r(x_2)\,dx_1\,dx_2 +$$

$$+ O(a_n^{-2}) = \frac{1}{2}[\int\int f^2(x_1)L_n^2(x_1, x_2)r(x_1)r(x_2)\,dx_1\,dx_2 +$$

$$+ \int\int f(x_1)f(x_2)L_n^2(x_1, x_2)r(x_1)r(x_2)\,dx_1\,dx_2] =$$

$$= 1/2(A_1 + A_2).$$

However

$$A_1 = 2a_n^{-1}\int f^2(x)r^2(x)\,dx \int K_0^2(x)\,dx + o(a_n^{-1})$$

and

$$A_2 = 2a_n^{-1}\int f^2(x)r^2(x)\,dx \int K_0^2(x)\,dx + o(a_n^{-1}).$$

Thus

$$DT_n^{(2)} = 2a_n^{-1}\int f^2(x)r^2(x)\,dx \int K_0^2(x)\,dx + o(a_n^{-1}). \tag{4.11}$$

Analogous calculation for the cumulants $\chi_n(m)$, $m > 2$, presented in Section 2 of Chapter 3 show that the m-th cumulant up to precision of terms of a higher order of smallness is

$$(m - 1)!1/2(a_n^{-1})^{m-1}[K*K]^{(m)}(0)\int f^m(x)r^m(x)\,dx. \tag{4.12}$$

From (4.7), (4.10), (4.11) and (4.12) we obtain the following basic

THEOREM 4.1. Under our assumptions on $f(x)$ and $K(x)$ and for $a_n^2n^{-1}\log^2 n = o(1)$ the random variable

$$a_n^{1/2}(T_n - \Theta_1(f) \int K^2(x)\, dx)$$

is distributed asymptotically normal with mean 0 and variance

$$\sigma^2 = 2\Theta_2(f) \int K_0^2(x)\, dx$$

where

$$\Theta_1(f) = \int f(x)r(x)\, dx, \quad \Theta_2 = \int f^2(x)r^2(x)\, dx.$$

From Theorem 4.1 follows

COROLLARY. Under the conditions of Theorem 4.1 the random variable

$$a_n^{1/2}(T_n - \Theta_1(f_n) \int K^2(x)\, dx)\sigma_n^{-1}$$

is distributed asymptotically normally (0, 1) where

$$\sigma_n^2 = 2\Theta_2(f_n) \int K_0^2(x)\, dx.$$

<u>Proof</u>. Setting $B_n^{(1)} = a_n^{1/2}[\Theta_1(f_n) - \Theta_1(f)]$ and $B_n^{(2)} = [\Theta_2(f_n) - \Theta_2(f)]$ it is sufficient to verify that $B_n^{(1)} = o_p(1)$ and $B_n^{(2)} = o_p(1)$. We have

$$E|B_n^{(1)}| \leq C_3 a_n^{1/2}\{\int E(f_n(x) - Ef_n(x))^2 r(x)\, dx\}^{1/2} +$$

$$+ O(a_n^{-3/2}) = O(a_n n^{-1/2}).$$

Hence,

$$B_n^{(1)} = O_p(1).$$

Furthermore

$$B_n^{(2)} \leq C_4 \sup_x | f_n(x) - Ef_n(x)| + O(a_n^{-2}).$$

However in view of (2.1.5) $\sup_x |f_n(x) - Ef_n(x)| = O_p(1)$. Thus $B_n^{(2)} = O_p(1)$.

This corollary allows us to construct an asymptotic test at level α for testing the hypothesis $H_0: f(x) = f(-x)$ for all $x \in (-\infty, \infty)$; critical regions for testing this hypothesis are established on the basis of the inequalities $T_n \geq \tilde{d}_n(\alpha)$:

$$\tilde{d}_n(x) = \Theta_1(f_n) \int K^2(x)\, dx + a_n^{-1/2}\lambda_\alpha \sigma_n,$$

where λ_α is the α-th percentile of the standard normal distribution.

3. Consider the local behavior of power of the same test in the case of two types of sequences of alternatives 'close' to the hypothesis $H_0: f(x) = f(-x)$. The first type of local alternatives is of the form

$$H_1: g_n(x) = f(x) + \alpha_n \phi(x), \quad \phi(x) \neq \phi(-x), \tag{4.13}$$

$$\int \phi(x)\, dx = 0 \quad \text{and} \quad \alpha_n \downarrow 0 \quad \text{as} \quad n \to \infty$$

THEOREM 4.2. Let $K(x) \in G$ and $g_n(x) \in \Pi$. If $a_n = n^{\delta}$ and $\alpha_n = n^{-1/2+\delta/4}$, $0 < \delta < 1/2$, then as $n \to \infty$

$$P_{H_1}\{T_n \geq \tilde{d}_n(\alpha)\} \to 1 - \Phi(\lambda_\alpha - \sigma^{-1}) \int [\phi(x) - \phi(-x)]^2 r(x)\, dx.$$

The proof of the theorem is analogous to the proof of Theorem 3.2.3 and of its corollary. We only note that for positive weight functions the test for testing H_0 - the symmetry of the distribution with respect to point $x = 0$ - against the alternatives of the form (4.13) is asymptotically strictly unbiased since $\int (\phi(x) - \phi(-x))^2 r(x)\, dx > 0$ and equals 0 if and only if $\phi(x) = \phi(-x)$, $x \in (-\infty, \infty)$.

Now consider a sequence of '<u>singular</u>' alternatives

$$H_1: g_n(x) = f(x) + \alpha_n \phi_n(x), \quad \phi_n(x) \quad \phi_n(-x),$$

where $\alpha_n \downarrow 0$ in an appropriate manner and $\phi_n(x)$ satisfies the conditions of Theorem 3.3.1. For simplicity we shall discuss the particular case (3.3.7), i.e. the case when $\phi_n(x) = W(x/\beta_n)$. However, Theorem 4.3 proven below is valid also in the general case.

THEOREM 4.3. Let $K(x) \in G$ and $g_n(x) \in \Pi$ and moreover, let $W''(x) \in L_2(-\infty, \infty)$.
If

$$n^{-1} a_n^2 \log^2 n \to 0, \quad \alpha_n = a_n^{\alpha} n^{-1/2}, \quad \beta_n = a_n^{-\beta}, \quad \beta > \alpha,$$

$$1/2 + \beta = 2\alpha, \quad 0 < \beta < 1$$

and $r(x)$ is continuous at point $x = 0$, $r(0) \neq 0$, then

$$P_{H_1}\{T_n \geq \tilde{d}_n(\alpha)\} \to 1 - \Phi\left(\lambda_\alpha - \frac{r(0)}{\sigma} \int [W(x) - W(-x)]^2\, dx\right)$$

as n increases.

<u>Proof</u>. Consider $P_{H_1}\{T_n \geq d_n(\alpha)\}$ where $d_n(\alpha)$ is obtained from $\tilde{d}_n(\alpha)$ by replacing $\Theta_1(f_n)$ and $\Theta_2(f_n)$ by $\Theta_1(f)$ and $\Theta_2(f)$ respectively. Tracing down the proof of Theorems 4.1 and 3.3.1 we obtain that as $n \to \infty$

$$P_{H_1}(T_n \geq d_n(\alpha)) \to 1 - \Phi\left(\lambda_\alpha - \frac{r(0)}{\sigma} \int [W(x) - W(-x)]^2\, dx\right).$$

Furthermore, since

$$a_n^{1/2} \sigma_n^{-1} (T_n - \Theta_1(f_n) \int K^2(x)\, dx) = a_n^{1/2} \sigma^{-1} (T_n - \Theta_1(f)) \frac{\sigma}{\sigma_n} +$$

$$+ a_n^{1/2} \sigma_n^{-1} (\Theta_1(f_n) - \Theta_1(f)) \int K^2(x)\, dx,$$

$$a_n^{1/2} E(|\Theta_1(f_n) - \Theta_1(f)|/H_1) = O\left(\frac{a_n}{\sqrt{n}}\right)$$

and

$$\Theta_2(f_n) - \Theta_2(f) = o_p(1),$$

the proof of the theorem follows immediately.

Remark. Integrating $g_n(x)$ we verify that alternatives differ from the null hypothesis by the quantity of the order $\alpha_n\beta_n = O(n^{-1/2})$. Thus the tests based on the deviation between empirical distribution functions, for example, tests of the ω_n^2-type and tests of the Kolmogorov-Smirnov type cannot distinguish the 'singular' hypothesis from the null hypothesis. Hence in view of Theorem 4.3 tests based on T_n for 'singular' alternatives are more powerful than the tests of the type mentioned above. Our tests is also asymptotically strictly unbiased since $\int (W(x) - W(-x))^2 dx > 0$ and equals 0 if and only if $W(x) = W(-x)$ almost everywhere in the Lebesgue measure.

5. Independence of Tests Based on Kernel-Type Density Estimators

1. Let $X_i = (X_i^{(1)}, X_i^{(2)})$, $i = 1, 2, \ldots$ be a sequence of independent identically distributed random variables with values in the Euclidean two-dimensional space R_2 possessing the density $f(\vec{x}) = f_1(x_1)f_2(x_2)$ (hypothesis H_0), $\vec{x} = (x_1, x_2)$ where $f_1(x_1)$ and $f_2(x_2)$ are marginal densities.

Consider the class of density estimators of the types (2.2.1) and (1.1.1) for the densities $f(\vec{x})$, $f_1(x_1)$, $f_2(x_2)$ respectively:

$$f_n(\vec{x}) = a_n^2 \int K(a_n(\vec{x} - \vec{u}))\, dF_n(u),$$

$$f_{n1}(x_1) = a_n \int K_1(a_n(x_1 - u_1))\, dF_{n1}(u_1),$$

$$f_{n2}(x_2) = a_n \int K_2(a_n(x_2 - u_2))\, dF_{n2}(u_2),$$

where $F_n(\vec{x})$ is the two dimensional sample empirical function, $F_{n1}(x_1)$ and $F_{n2}(\vec{x}_2)$ are the corresponding marginal distribution functions, $K(x) = K_1(x_1)\cdot K_2(x_2)$ and $K_1(x_1)$ and $K_2(x_2)$ satisfy conditions G, while $f_1(x_1)$ and $f_2(x_2)$ belong to the class of densities Π (cf. Section 2 of Chapter 3). We choose the quantity

$$S_n = na_n^{-2} \int [f_n(\vec{x}) - f_{n1}(x_1)\cdot f_{n2}(x_2)]^2 r(\vec{x})\, d\vec{x},$$

where $r(\vec{x}) = r_1(x_1)\cdot r_2(x_2)$, $r_1(x_1)$ and $r_2(x_2)$ are piecewise continuous bounded and integrable functions as a measure of deviation between $f_n(\vec{x})$ and $f_{n1}(x_1)\cdot f_{n2}(x_2)$.

In this section the problem of the limiting distribution of the functional S_n is solved and a test for testing the hypothesis $H_0: f(\vec{x}) = f_1(x_1)\cdot f_2(x_2)$ is constructed.

2. Introduce the empirical process

$$T_n(\vec{x}) = \sqrt{n}\,[F_n(\vec{x}) - F_{n1}(x_1)\cdot F_{n2}(x_2)]$$

and the two-dimensional Brownian bridge $W^0(\vec{y})$, $\vec{y}$ $\Delta = [0, 1]\times[0, 1]$, i.e.

$$W^0(\vec{y}) = W(\vec{y}) - y_1 y_2 W(1, 1).$$

Here $W(\vec{y})$ is the standard two-dimensional Wiener random field on Δ: $W(\vec{y})$ is a separable Gaussian process with $EW(\vec{y}) = 0$, $\operatorname{cov} W(\vec{x})W(\vec{y}) = \min(x_1, y_1)\cdot\min(x_2, y_2)$, $\vec{x} = (x_1, x_2)$, $\vec{y} = (y_1, y_2)$.

Our subsequent proofs are based on the following theorem.

THEOREM 5.1 ([55]). Let $F(x_1, x_2) = F_1(x_1)\cdot F_2(x_2)$ where $F(x_1, x_2)$ is the distribution function of X_1 while $F_1(x_1)$ and $F_2(x_2)$ are the marginal distribution functions of $X_1^{(1)}$ and $X_1^{(2)}$ respectively. Then there exists a sequence of two dimensional Brownian bridges $\{W_n^0(\vec{x}),\ \vec{x} \in \Delta\}$ such that with probability one

$$\sup_{x\in\Delta} |T_n(F_1^{-1}(x_1),\quad F_2^{-1}(x_2)) - T^{(n)}(x_1, x_2)| = O_p(n^{-1/2}\log^2 n),$$

where

$$T^{(n)}(\vec{x}) = W_n^0(\vec{x}) - W_n^0(x_1, 1)x_2 - W_n^0(1, x_2)x_1, \quad \vec{x} = (x_1, x_2) \in \Delta.$$

Next consider the random processes:

$$Y_n(\vec{x}) = \sqrt{n}\, a_n^{-1}[f_n(\vec{x}) - f_{n1}(x_1)f_{n2}(x_2)] =$$

$$= a_n^{-1}\int K_1(a_n(x_1 - u))K_2(a_n(x_2 - v))\, dT_n(u, v)$$

and

$$Z_n(\vec{x}) = a_n^{-1}\int K_1(a_n(x_1 - u))K_2(a_n(x_2 - v))\, dT^{(n)}(F_1(u), F_2(v)).$$

Set

$$\xi_n = \int Z_n^2(\vec{x})r(\vec{x})\, d\vec{x}.$$

LEMMA 5.1. $\sup_{\vec{x}\in R_2} |Y_n(\vec{x}) - Z_n(\vec{x})| = O_p(n^{-1/2}a_n \log^2 n)$.

Proof. We have:

$$\sup_{\vec{x} \in R_2} |Y_n(\vec{x}) - Z_n(\vec{x})| = a_n^{-1} \sup_{\vec{x} \in R_2} |\int K_1(a_n(x_1 - u)) \times$$

$$\times K_2(a_n(x_2 - v)) \, d(T_n(u, v) - T^{(n)}(F_1(u), F_2(v)))| \leq$$

$$\leq a_n \sup_{(u,v) \in \Delta} |T_n(u, v) - T^{(n)}(F_1(u),$$

$$F_2(v))| \bigvee_{-\infty}^{\infty} (K_1) \cdot \bigvee_{-\infty}^{\infty} (K_2)$$

Whence, utilizing Theorem 5.1 we verify the validity of the lemma.

LEMMA 5.2. The random variable $a_n \sigma^{-1}(f)(\xi_n - A(f, n))$ is asymptotically normal with mean 0 and variance 1, where

$$A(f, n) = \int f(\vec{x}) r(\vec{x}) \, d\vec{x} \int K^2(\vec{u}) \, d\vec{u} - a_n^{-1} \int K_1^2(u) \, du \int f_1(x_1) \times$$

$$\times f_2^2(x_2) r(\vec{x}) \, d\vec{x} - a_n^{-1} \int K_2^2(u) \, du \int f_2(x_2) f_1^2(x_1 r(\vec{x}) \, d\vec{x},$$

$$\sigma^2(f) = 2 \int f^2(\vec{u}) r^2(\vec{u}) \, d\vec{u} \int (K_1 * K_1(x))^2 \, dx \times$$

$$\times \int (K_2 * K_2(x))^2 \, dx.$$

Proof. The covariance function $R_n(\vec{x}, \vec{y})$ of the process $Z_n(\vec{x})$ is of the form

$$R_n(\vec{x}, \vec{y}) = a_n^2 [\int K_1(a_n(x_1 - u)) K_1(a_n(y_1 - u)) K_2(a_n(x_2 - v)) \times$$

$$\times K_2(a_n(y_2 - v) f(u, v) \, du \, dv - \int K_1(a_n(x_1 - u)) \times$$

$$\times K_1(a_n(y_1 - u)) f_1(u) \, du \int K_2(a_n(x_2 - v)) f_2 \, (v) \, dv \times$$

$$\times \int K_2(a_n(y_2 - v)) f_2(v) \, dv - \int K_2(a_n(x_2 - v)) K_2 \times$$

$$\times (a_n(y_2 - v)) f_2(v) \, dv \int K_1(a_n(x_1 - u) f_1(u) \, du \times$$

$$\times \int K_1(a_n(y_1 - u)) f_1(u) \, du + \int K(a_n(\vec{x} - \vec{u})) K \times$$

$$\times (a_n(\vec{y} - \vec{v})) f(\vec{u}) f(\vec{v}) \, d\vec{u} \, d\vec{v}] = [f_1(x_1) K_1 * K_1(a_n(y_1 -$$

$$x_1)) + O(a_n^{-1})][f_2(x_2) K_2 * K_2(a_n(Y_2 - x_2)) + O(a_n^{-1})]. \tag{5.1}$$

The k-th order cumulant $\chi_n(k)$ of the random variable ξ_n is given by the formula

$$\chi_n(k) = (k-1)!2^{k-1} \int \dots \int R_n(\vec{x}_1, \vec{x}_2) R_n(\vec{x}_2, \vec{x}_3) \dots$$
$$\dots R_n(\vec{x}_k, \vec{x}_1) r(\vec{x}_1) \dots r(\vec{x}_k)\, d\vec{x}_1 \dots d\vec{x}_k, \tag{5.2}$$

which follows from the representation of the characteristic function of the square of the L_2-norm of a Gaussian process (cf. [53] p. 417 or [49]).

It is easy to calculate that

$$E\xi_n = A(f, n) + O(a_n^{-2}) \quad \text{and} \quad \lim_{n\to\infty} a_n^2 D\xi_n = \lim_{n\to\infty} a_n^2\, \chi_n(2) = \sigma^2(f).$$

Simple calculations utilizing (5.1) and (5.2) show (cf. Section 2 of Chapter 3 that $\chi_n(k)$ up to the terms of higher order of smallness is equal to

$$(k-1)!2^{(k-1)} a_n^{-2(k-1)} [K_1 * K_1]^{(k)}(0) [K_2 * K_2]^{(k)}(0) \int f^k(\vec{x}) r^k(\vec{x})\, d\vec{x}.$$

The lemma is thus proved.

THEOREM 5.2. Under our assumption on $f(\vec{x})$ and $K(\vec{x})$, $\vec{x} = (x_1, x_2)$ and for $a_n = n^\alpha$, $0 < \alpha < 1/3$, the random variable

$$a_n \sigma^{-1}(f)(S_n - A(f, n))$$

is asymptotically normal with mean 0 and variance 1.

Proof. Utilizing expression (5.1) we verify that

$$\int R_n(\vec{x}, \vec{y}) r(\vec{x}) r(\vec{y})\, d\vec{x}\, d\vec{y} = O(a_n^{-2}),$$

and thus

$$E \left| \int Z_n(\vec{x}) r(\vec{x})\, d\vec{x} \right| = O(a_n^{-1}).$$

Whence applying Lemma 5.1 we obtain

$$S_n = \int Z_n^2(\vec{x}) r(\vec{x})\, d\vec{x} + O_p(n^{-1} a_n^2 \log^4 n).$$

Thus

$$a_n \sigma^{-1}(f)(S_n - \xi_n) = O_p(n^{-1} a_n^3 \log^4 n),$$

i.e. this expression converges in probability to zero for our choice of a_n. The assertion of the theorem now follows from Lemma 5.2.

In connection with Theorem 5.2 it should be noted that the statistic S_n is normalized by the quantities $A(f_1, f_2, n)$ and $\sigma(f_1, f_2)$ which depend on the unknown probability density of observations. Therefore for applications it is necessary to prove the analog of Theorem 5.2 where $A(f_1, f_2, n)$ and $\sigma(f_1, f_2)$ are replaced by $A_n = A(f_{n_1}, f_{n_2})$ and $\sigma_n = \sigma(f_{n_1}, f_{n_2})$ respectively. Thus the following theorem is valid.

THEOREM 5.3. Under the conditions of Theorem 5.2 the random variable $a_n\sigma_n^{-1}(S_n - A_n)$ is asymptotically normal with mean 0 and variance 1.

Proof. Clearly it is sufficient to prove that

$$1°. \quad a_n(A_n - A) = o_p(1) \quad \text{and} \quad 2°. \quad \sigma_n^2 - \sigma^2 = o_p(1).$$

We have

$$a_n(A_n - A) = a_n \int (f_1(x_1)f_2(x_2) - f_{n1}(x_1)f_{n2}(x_2))r(\vec{x})\, d\vec{x} +$$

$$+ \int [f_1(x_1)f_2^2(x_2) - f_{n1}(x_1)f_{n2}^2(x_2)]r(\vec{x})\, d\vec{x} +$$

$$+ \int [f_2(x_2)f_1^2(x_2) - f_{n2}(x_1)f_{n1}^2(x)]r(\vec{x})\, d\vec{x} =$$

$$= I_1 + I_2 + I_3.$$

Note that

$$I_1 = a_n \int [Ef_{n1}(x_1)Ef_{n2}(x_2) - f_{n1}(x_1)f_{n2}(x_2)]r(\vec{x})\, d\vec{x} +$$

$$+ O(a_n^{-1}) = I_n^{(1)} + O(a_n^{-1}).$$

On the other hand

$$E|I_n^{(1)}| \le C_1 a_n \{ \int [E(f_{n1}(x_1) - Ef_{n1}(x_1))^2 \cdot E(f_{n2}(x_2) -$$

$$- Ef_{n2}(x_2))^2 r(\vec{x})\, d\vec{x}\}^{1/2} \le C_2 n^{-1/2} a_n^{3/2} = o(1).$$

Consequently,

$$I_1 = o_p(1). \tag{5.3}$$

Now consider I_2. It is easy to verify that

$$I_2 \le \sup_{-\infty<x<\infty} |f_{n1}(x) - f_1(x)| \int f_{n2}^2(x_2)r(\vec{x})\, d\vec{x} +$$

$$+ 2 \sup_{-\infty<x<\infty} | f_{n2}(x) - f_2(x)|.$$

Whence utilizing Theorem 2.1 and relationship 2.2.1 we obtain that

$$I_2 = o_p(1). \tag{5.4}$$

Analogously,

$$I_3 = o_p(1) \tag{5.5}$$

Thus relations (5.3)-(5.5) imply 1°. From Theorems 2.1.1 and 2.2.1 we immediately arrive at 2°. The theorem is thus proved.

Theorem 5.3 allows us to construct an asymptotic test of level α for testing the hypothesis $H_0: f(x_1, x_2) = f_1(x_1) \cdot f_2(x_2)$; critical regions for testing this hypothesis are determined by the inequality $S_n \geq d_n(\alpha)$:

$$d_n(\alpha) = A(f_{n1}, f_{n2}, n) + a_n^{-1} \lambda_\alpha \sigma_n,$$

where λ_α is the α-th percentile of the standard normal distribution.

Remark to Theorem 5.2. In paper [56] an analogous theorem was proved for the case when the kernel $K(\vec{x})$ belongs to some narrow class of function; moreover, in this theorem the normalizing quantity $A(f_1, f_2, n)$ is stated somewhat inaccurately.

Chapter 4

NONPARAMETRIC ESTIMATION OF REGRESSION CURVES AND COMPONENTS OF A CONVOLUTION

1. Some Asymptotic Properties of Nonparametric Estimators of Regression Curves

In this chapter asymptotic properties of certain functions of non parametric estimators of a probability density of types 1.1.1 and 1.2.1 are studied. In particular in this and in the succeeding three sections we shall investigate properties of the non parametric estimators $r_n(x)$ of the regression curve $r(x)$ which was introduced and studied by the author in [21]. Interest in this estimator has been constantly growing and many interesting results were obtained in the USSR as well as abroad (cf., for example, [10], [16], [22], [23], etc.).

1. Let (X, Y) be a two-dimensional random variable with the density function $f(x, y)$ and $g(x)$ is the marginal density function of the random variable X. Let $E|Y| < \infty$ and denote the regression function of Y on X by $r(x)$, i.e. $r(x) = E(Y|X = x)$. Select a sample (X_i, Y_i), $i = \overline{1, n}$ from the population (X, Y) and as an approximation to the regression curve $r(x)$ consider the class of statistics

$$r_n(x) = \begin{cases} \dfrac{\sum_{i=1}^{n} Y_i K(a_n(x - X_i))}{\sum_{i=1}^{n} K(a_n(x - X_i))} & \text{if } \sum_{i=1}^{n} K(a_n(x - X_i)) \neq 0 \\ 0 & \text{if } \sum_{i=1}^{n} K(a_n(x - X_i)) = 0. \end{cases}$$

The kernel $K(x)$ is an arbitrary density function satisfying the conditions:

1°. $\sup_{-\infty<x<\infty} K(x) < \infty$,

2°. $\lim_{|x|\to\infty} |x|K(x) = 0$.

3°. $K(x) = K(-x)$,

4°. $x^2K(x) \in L_1(-\infty, \infty)$.

Denote

$$\phi_n(x) = n^{-1}a_n \sum_{i=1}^{n} Y_i K(a_n(x - X_i)),$$

$$g_n(x) = n^{-1}a_n \sum_{i=1}^{n} K(a_n(x - X_i)),$$

$$\phi(x) = \int y f(x, y)\, dy, \qquad V(x) = \int y^2 f(x, y)\, dy.$$

THEOREM 1.1. Let $EY^2 < \infty$ and $na_n^{-1} \to \infty$. Then for any x for which $r(x)$, $g(x)$ and $V(x)$ are continuous and $g(x) > 0$, $r_n(x)$ is a consistent estimator of $r(x)$.

Proof. Since $g_n(x)$ is a consistent estimator of $g(x)$ it is sufficient to verify that $\phi_n(x)$ is a consistent estimator of $\phi(x)$.

We have

$$E\phi_n(x) = a_n E(YK(a_n(x - X))) = \iint a_n y K(a_n(x - t)) f(t, y)\, dt\, dy =$$
$$= a_n \int E[YK(a_n(x - X)) | X = t] g(t)\, dt = a_n \int K(a_n(x -$$
$$- t)) r(t) g(t)\, dt = a_n \int K(a_n(x - t))\phi(t)\, dt. \tag{1.1}$$

Since $\phi(x)$ is integrable and continuous at point x in view of Theorem IA in [5] it follows from (1.1) that

$$E\phi_n(x) \to \phi(x) \quad \text{as} \quad n \to \infty. \tag{1.2}$$

By the same theorem we obtain

$$\bar{D}\phi_n(x) = n^{-1}a_n^2 \int K^2(a_n(x - t)) V(t)\, dt - n^{-1}(E\phi_n(x))^2 \sim$$
$$\sim n^{-1}a_n V(x) \int K^2(t)\, dt. \tag{1.3}$$

Relations (1.2) and (1.3) imply that $E(\phi_n(x) - \phi(x))^2 \to 0$ as $n \to \infty$. The theorem is thus proved.

THEOREM 1.2. a) If $|Y| \leq C_1 < \infty$ a.s. and $na_n^{-1} \to \infty$ then

$$Er_n(x) = E\phi_n(x)/Eg_n(x) + O(n^{-1}a_n);$$

b) if $EY^2 < \infty$ and $na_n^{-2} \to \infty$ then

$$Er_n(x) = E\phi_n(x)/Eg_n(x) + O(n^{-1/2}a_n).$$

Assertions a) and b) imply that $r_n(x)$ is an asymptotically unbiased estimator of $r(x)$ for any x at which $\phi(x)$, $g(x)$ and $V(x)$ are continuous.

Proof. Utilizing the identity

$$1/u = 1/C - C^{-2}(u - C) + C^{-2}u^{-1}(u - C)^2, \qquad C \neq 0,$$

we obtain

$$Er_n(x) = E\phi_n(x)/Eg_n(x) - (Eg_n(x))^{-2}E[\phi_n(x) - E\phi_n(x)][g_n(x) -$$

$$- Eg_n(x)] + E\{g_n^{-1}(x)(Eg_n(x))^{-2}\phi_n(x)(g_n(x) - Eg_n(x))^2\} =$$

$$= E\phi_n(x)/Eg_n(x) + [C_n^{(1)}(x) + C_n^{(2)}(x)](Eg_n(x))^{-2}. \quad (1.4)$$

We shall first prove a). Note that $Dg_n(x) \sim n^{-1}a_n g(x) \int K^2(x)\, dx$. This together with (1.3) implies that

$$C_n^{(1)}(x) = O(n^{-1}a_n). \quad 1.5)$$

Since the distribution of Y is bounded we obtain

$$|C_n^{(2)}(x)| \leq C_1 E[g_n(x) - Eg_n(x)]^2 \sim C_1 n^{-1} a_n g(x) \int K^2(u)\, du. \quad (1.6)$$

Relations (1.4), (1.5) and (1.6) yield assertion a).

To prove b) observe that relation (1.5) remains valid also in our case b) i.e. when the component Y is distributed along the whole axis.

Next,

$$|C_n^{(2)}(x)| \leq E \max\{|Y_1|, \ldots, |Y_n|\}|g_n(x) - Eg_n(x)|^2 \leq$$

$$\leq \left(\sum_{i=1}^{n} EY_i^2\right)^{1/2} [E(g_n(x) - Eg_n(x)]^4)^{1/2} =$$

$$= \sqrt{n}\,(EY^2)^{1/2}[E(g_n(x) - Eg_n(x)]^4)^{1/2}.$$

In view of Lemma 1.2.3, $E(g_n(x) - Eg_n(x))^4 = O(n^{-2}a_n^2)$. Consequently

$$C_n^{(2)}(x) = O(n^{-1/2}a_n). \quad (1.7)$$

Relations (1.4), (1.5) and (1.7) yield assertion b).

THEOREM 1.3. Let $E|Y|^{2+\delta} < \infty$ for some $\delta \in (1/2, 1]$. Next let $V(x)$ be continuous, $\phi(x)$ and $g(x)$ possess boundeu derivatives up to the second order. Then if $n^\delta a_n^{-2-\delta} \to \infty$, $na_n^{-5} \to 0$, we obtain that

$$\sqrt{\frac{n}{a_n}}\,(r_n(x) - r(x))$$

is asymptotically distributed normally $(0, \sigma(x))$ where

$$\sigma^2(x) = D(Y|X = x)/g(x) \cdot \int K^2(u)\, du.$$

Proof. We have

$$r_n(x) - E\phi_n(x)/Eg_n(x) = \zeta_n(x) g_n^{-1}(x)(Eg_n(x))^{-1}, \quad (1.8)$$

where

$$\zeta_n(x) = n^{-1}a_n \sum_{i=1}^{n} (\alpha_n Y_i - \beta_n)K(a_n(x - X_i)),$$

and moreover

$$E\zeta_n(x) = 0, \quad \alpha_n = Eg_n(x) \to g(x) \quad \text{and} \quad \beta_n = E\phi_n(x) \to \psi(x).$$

It is easy to verify that

$$na_n^{-1} D\zeta_n(x) = \alpha_n^2 a_n \int K^2(a_n(x - t))V(t)g(t)\, dt - 2\alpha_n\beta_n a_n \int K^2 \times$$

$$\times (a_n(x - t))\phi(t)\, dt + \beta_n^2 a_n \int K^2(a_n(x - t))g(t)\, dt.$$

Whence in view of the Theorem IA [5] we have

$$na_n^{-1} D\zeta_n(x) \to \nu(x) = \vec{D}(Y|X = x)g^3(x) \int K^2(u)\, du. \tag{1.9}$$

We shall now prove that $\zeta_n(x)/\sqrt{D\zeta_n(x)}$ is distributed asymptotically normally (0, 1). For this purpose, it is sufficient to show that

$$nP\left\{\left|\frac{z_n}{\sqrt{Dz_n}}\right| \geq \varepsilon n^{1/2}\right\} \to 0 \tag{1.10}$$

as $n \to \infty$ where

$$z_n = a_n(\alpha_n Y - \beta_n)K(a_n(x - X)).$$

To verify (1.10) it is sufficient to show that

$$L_n = \frac{E|z_n|^{2+\delta}}{n^{\delta/2}(Dz_n)^{1+\delta/2}} \to 0. \tag{1.11}$$

Condition (1.11) is fulfilled. Indeed

$$E|z_n|^{2+\delta} \leq a_n^{2+\delta}\alpha_n^{2+\delta}\iint |y|^{2+\delta}K^{2+\delta}(a_n(x - t))f(x, t)\, dx\, dt + \beta_n^{2+\delta} \times$$

$$\times a_n^{2+\delta} \int\int K^{2+\delta}(a_n(x - u))f(u, y)\, du\, dy \leq C_2 E|Y|^{2+\delta} a_n^{2+\delta}$$

and

$$(a_n^{-1} Dz_n)^{1+\delta/2} \to (\nu(x))^{1+\delta/2}.$$

Hence

$$L_n = O(n^{-\delta/2}a_n^{1+\delta/2}).$$

It follows from (1.9) and the preceding assertion that

$$\sqrt{na_n^{-1}}\, \nu^{-1/2}(x)\zeta_n(x)$$

is distributed asymptotically normally (0,1). Whence, using (1.8) and taking into account that $g_n(x)Eg_n(x)$ converges in probability to $g^2(x)$, we obtain that

$$P\{\sqrt{na_n^{-1}}\,\sigma^{-1}(x)(r_n(x) - E\phi_n(x)/Eg_n(x)) < \lambda\} \to \Phi(\lambda). \tag{1.12}$$

We now use (1.1) to arrive at

$$E\phi_n(x) = \phi(x) + 1/2a_n^{-2} \int \phi''(x + \Theta u a_n^{-1})u^2K(u)\,du =$$
$$= \phi(x) + O(a_n^{-2}), \quad 0 < \Theta < 1. \tag{1.13}$$

Analogously,

$$Eg_n(x) = g(x) + O(a_n^{-2}). \tag{1.14}$$

Thus (1.12), (1.13) and (1.14) yield the validity of the theorem

THEOREM 1.4. Let $|Y| \le C_3$ a.s., $V(x)$ be continuous, $\phi(x)$ and $g(x)$ possess bounded first order derivatives. Then:

a) if $a_n/n \to 0$ then

$$r_n(x) - r(x) = g^{-1}(x)[\eta_n(x) - E\eta_n(x)] + O_p\left(a_n^{-1} + \frac{a_n}{n}\right), \tag{1.15}$$

where

$$\eta_n(x) = 1/n \sum_{i=1}^{n} \eta_n^{(i)}(x), \quad \eta_n^{(i)}(x) = a_n(Y_i - r(x))K(a_n(x - X_i)); \tag{1.16}$$

b) if $na_n^{-3} \to 0$ and $a_n/n \to 0$, then

$$\sqrt{\frac{n}{a_n}}\,(r_n(x) - r(x))$$

is distributed asymptotically normally $(0, \sigma(x))$.

<u>Proof</u>. Set $\overline{r_n}(x) = r_n(x) - r(x)$. Evidently $\eta_n(x) = \phi_n(x) - r(x)g_n(x)$ so that

$$E|\overline{r_n}(x) - g^{-1}(x)\eta_n(x)| = E|\overline{r_n}(x)(g_n(x) - g(x))g^{-1}(x)|.$$

Therefore

$$E|\overline{r_n}(x) - g^{-1}(x)\eta_n(x)| \le 1/2g^{-1}(x)[E(r_n(x) - r(x))^2 +$$
$$+ E(g_n(x) - g(x))^2]. \tag{1.17}$$

The identity

$$r_n(x) = \phi_n(x)/g(x) - \phi_n(x)(g_n(x) - g(x))g^{-1}(x)g_n^{-1}(x)$$

implies that

$$E(r_n(x) - r(x))^2 = E[\phi_n(x) - \phi(x) - \phi_n(x)g_n^{-1}(x)(g_n(x) -$$
$$- g(x))]^2 g^{-2}(x).$$

Whence taking the inequality $\phi_n(x)/g_n(x) \leq C_3$ a.s. into account (the latter follows from the fact that Y is bounded) we obtain

$$E(r_n(x) - r(x))^2 \leq C_4 g^{-2}(x)(E(\phi_n(x) - \phi(x))^2 + E(g_n(x) - g(x))^2).$$

We now estimate $E(\phi_n(x) - \phi(x))^2$. The continuity and integrability of V(x) imply that

$$\bar{D}\phi_n(x) \leq a_n/n \int K^2(a_n(x - u))V(u)\, du = O(a_n/n).$$

Next, the condition on $\phi(x)$ yields that $E\phi_n(x) = \phi(x) + O(a_n^{-1})$. Hence $E(\phi_n(x) - \phi(x))^2 = O(a/n + a_n^{-2})$. Analogously it can be shown that $E(g_n(x) - g(x))^2 = O(a_n/n + a_n^{-2})$. Whence $E(r_n(x) - r(x))^2 = O(a_n^{-2} + a_n/n)$. In view of these bounds and (1.17) we arrive at

$$r_n(x) - r(x) = g^{-1}(x)\eta_n(x) + O_p(a_n^{-2} + a_n/n). \tag{1.18}$$

Finally to complete the proof of the theorem it remains only to note the bound $E\eta_n(x) = O(a_n^{-1})$ in (1.18). Assertion a) is thus proved.

Assertion b) follows directly from (1.15) since $(\eta_n(x) - E\eta_n(x)) \times (\vec{D}\eta_n(x))^{-1/2}$ is distributed asymptotically normally $(0, \sigma(x)\cdot g(x))$.

COROLLARY TO THEOREM 1.4. Let t_j, $t_j \neq t_i$, $j = \overline{1, m}$, $i = \overline{1, m}$ be fixed points such that $g(t_j) > 0$. It is easy to verify that $na_n^{-1} \operatorname{cov} \eta_n(t_j) \times \eta_n(t_i) = O(1)$. Then it follows from b) and the multidimensional central limit theorem that the random vector $(\sqrt{na_n^{-1}}(r_n(t_j) - r(t_j)), j = \overline{1, m})$ is distributed asymptotically normally with mean 0 and the covariance matric (σ_{ij}) where $\sigma_{ij} = 0$, $i \neq j$, $\sigma_{ii} = \sigma(t_i)$, $i = \overline{1, m}$.

Theorems 1.3 and 1.4 allow us to construct a confidence interval for r(x). For this purpose, the quantity $\sigma(x)$ should be replaced by a consistent estimator. For example, one can choose for this estimator the statistic

$$\sigma_n^2(x) = na_n^{-1} \frac{\sum_{i=1}^{n} (Y_i - r_n(x))^2 K(a_n(x - X_i))}{\left[\sum_{i=1}^{n} K(a_n(x - X_i))\right]^2} \int K^2(v)\, dv. \tag{1.19}$$

Then under the conditions of Theorems 1.3 and 1.4 we have

$$P\{|r_n(x) - r(x)| < \lambda\sigma_n(x)/\sqrt{na_n^{-1}}\} \to 2\Phi(\lambda) - 1 \quad \text{as} \quad n \to \infty. \tag{1.20}$$

2. The simplest way to characterize the error in estimating an unknown regression curve is to use the mathematical expectation of the quadratic deviation

$$U_n = E \int \eta_n^2(x)h(x)\, dx = E \int (r_n(x) - r(x))^2 g_n^2(x)h(x)\, dx, \quad (1.21)$$

where $h(x)$ is the weight function which we shall assume to be bounded and integrable. The quantity U_n defines the risk function of the estimator $r_n(x)$.

THEOREM 1.5. Let the second order derivatives of functions $\phi(x)$ and $g(x)$ be continuous, bounded and square integrable with the weight $h(x)$. If, moreover $\sup_{-\infty<x<\infty} |V(x)| < \infty$, $r(x)h(x)$ and $r^2(x)h(x) \in L_1(-\infty, \infty)$ then

$$U_n \sim A_1(f, K)a_n/n + A_2(f, K)a_n^{-4}, \quad (1.22)$$

where

$$A_1(f, K) = \int \vec{D}(Y\ X = x)g(x)h(x)\, dx \int K^2(u)\, du,$$

$$A_2(f, K) = 1/4 \int (\phi''(x) - g''(x)r(x))^2 h(x)\, dx (\int u^2 K(u)\, du)^2.$$

Proof. Fubini's theorem yields

$$U_n = \int E(Y - r(x))^2 K^2(a_n(x - X))h(x)\, dx \frac{a_n^2}{n} + (1 - 1/n)a_n^2 \times$$

$$\times \int [E(Y - r(x))K(a_n(x - X))]^2 h(x)\, dx = D_n + E_n.$$

Omitting the detailed proof in the course of which the Lebesgue theorem on the limiting transition under the integration sign is used repeatedly one can write that

$$D_n \sim n^{-1}a_n(\int V(x)h(x)\, dx - 2 \int \phi(x)r(x)h(x)\, dx + \int r^2(x)g(x) \times$$

$$\times h(x)\, dx) \int K^2(u)\, du = n^{-1}a_n \int D(Y|X = x)g(x)h(x)\, dx \times$$

$$\times \int K^2(u)\, du.$$

We shall now estimate E_n. Utilizing Taylor's theorem with the remainder in the integral form we have

$$E_n = a_n^{-4} \int [\int \int (1 - t)(\phi''(x + tua_n^{-1}) - r(x)g''(x + tua_n^{-1}))^2 \times$$

$$\times u^2 K(u)\, du\, dt]^2 h(x)\, dx = a_n^{-4} A_n.$$

By the Lebesgue theorem, $A_n \to A_2(f, K)$. Thus $E_n \sim A_2(f, K)a_n^{-4}$.

These asymptotic representations of D_n and E_n prove our theorem.

We shall now determine the asymptotic value $a_n = a_n^0$ which minimizes the r.h.s. of (1.22). The optimal value of a_n is equal to

$$a_n^0 = (4A_2(f, K)/A_1(f, K))^{1/5} n^{1/5}.$$

For this choice of a_n

$$U_n \sim (4^{1/5} + 4^{-4/5}) A_2^{1/5}(f, K) A_1^{4/5}(f, K) n^{-4/5}.$$

Remark. Theorem 1.5 admits a generalization. In particular if $\psi(x)$, $g(x) \in W_s$ and $K(x) \in H_s$ (see Section 1 of Chapter 1 for the definitions of W_s and H_s) then the optimal value of a_n is equal to

$$a_n^0 = C_1(f, K) n^{1/(1+2s)}$$

and for $a_n = a_n^0$

$$U_n \sim C_2(f, K) n^{-2s/(1+2s)},$$

where $C_1(f, K)$ and $C_2(f, K)$ are functionals which depend on $f(x, y)$ and $K(x)$.

2. Strong Consistency of Regression Curve Estimators in the Norm of the Space C(a, b)

In this section sufficient conditions for convergence in $C(a, b)$ a.s. of estimators of the regression curve to the theoretical $r(x)$ are presented.

We shall assume that $g(x)$ and $r(x)$ are continuous on the whole axis and that $K(x)$ in addition to conditions 1°-4° of the preceding section also satisfies condition 5°: $K(x)$ is a function with a bounded variation.

THEOREM 2.1. Let $K(x)$ satisfy the conditions 1°-5° with probability 1, $-\infty < A \leq Y \leq B < \infty$, $\min\limits_{-\infty<a\leq x\leq b<\infty} g(x) = \mu > 0$ and the series $\sum\limits_{n=1}^{n} e^{-\gamma n a_n^{-2}}$ is convergent for any $\gamma > 0$. Then as $n \to \infty$

$$\sup_{a\leq x\leq b} |r_n(x) - r(x)| \to 0 \quad \text{a.s.}$$

Without loss of generality we shall assume for simplicity that $A = 0$ and $B = 1$. The theorem is proved with the aid of the following two lemmas.

LEMMA 2.1. $\sup\limits_{a\leq x\leq b} |E\phi_n(x) - \phi(x)| \to 0$ as n increases.

Proof. Let $\delta > 0$. Then setting $M = \max\limits_{a\leq x\leq b} |g(x)r(x)|$ and taking (4.1.1) into consideration we have

$$\Big|\sup_{a\leq x\leq b} E\phi_n(x) - \phi(x)\Big| \leq \sup_{a\leq x\leq b} \sup_{|t|\leq\delta} |\phi(x - t) - \phi(x)| \int \times$$

$$\times K(x)\, dx + \frac{1}{\delta} \sup_{|t|\geq\delta/a_n^{-1}} |t| K(t) +$$

$$+ M \int_{|t|\geq\delta a_n} K(t)\, dt. \tag{2.1}$$

Let $\eta > 0$ be arbitrarily small. By choosing $\delta > 0$ sufficiently small the first summand on the r.h.s. can be made less than $\eta/2$. For the chosen $\delta > 0$ one can then select n so large that the second and third terms on the r.h.s. will also be less than $\eta/2$. Then it follows from (2.1) that

$$\sup_{a\leq x\leq b} |E\phi_n(x) - \phi(x)| < \eta.$$

The lemma is thus proved.

LEMMA 2.2. If K(x) satisfies conditions 1°-5° and $0 \leq Y \leq 1$ a.s. then

$$P\{\sup_{a\leq x\leq b} |\phi_n(x) - E\phi_n(x)| > \lambda\} \leq C_0 e^{-a_1 n a_n^{-2}} + C_1 e^{-a_2 n a_n^{-2}}.$$

<u>Proof</u>. $\phi_n(x)$ can be represented in the form

$$\phi_n(x) = a_n \int\int_0^1 vK(a_n(x - u))\, dS_n(u, v), \tag{2.2}$$

where $S_n(u, v)$ is a two-dimensional empirical distribution function.

Applying the formula of integration by parts for a two-dimensional integral ([44]) and (2.2) we arrive at

$$\sup_{a\leq x\leq b} |\phi_n(x) - E\phi_n(x)| = \sup_{a\leq x\leq b} |a_n \int [\vec{F}_1(u) - S_n^{(1)}(u)] \times$$
$$\times\, dK\, a_n(x - u)) + a_n \int\int_0^1 [F(u\cdot v) - S_n(u, v)] \times$$
$$\times\, dK\, a_n(x - u)\, dv| \leq \ell_1 a_n [\sup_{-\infty\leq u<\infty} |S_n^{(1)}(u) -$$
$$- F_1(u)| + \sup_{(u,v)\in R_2} |F(u, v) - S_n(u, v)|], \tag{2.3}$$

where F(u, v) is the distribution function of (X, Y), $F_1(u)$ is the marginal distribution function of X; $S_n^{(1)}(u)$ is the distribution function of the sample $X_1, X_2, \ldots, X_n$ and $\ell_1 = \overset{\infty}{\underset{-\infty}{V}}(K)$

Utilizing the inequalities (2.1.4) and (2.1.10) we obtain from (2.3) the bound

$$P\{\sup_{a\leq x\leq b} |\phi_n(x)-E\phi_n(x)|>\lambda\} \leq P\left\{\sup_{z\in R_1} |S_n^{(1)}(z) - F_1(z)| > \frac{\lambda}{2\ell_1 a_n}\right\} +$$
$$+ P\left\{\sup_{(u,v)\in R_2} |S_n(u, v) - F(u, v)| > \frac{\lambda}{2\ell_1 a_n}\right\} \leq$$
$$\leq C_3 e^{-\alpha_1 n a_n^{-2}} + C_4 e^{-\alpha_2 n a_n^{-2}},$$

where

$$\alpha_1 = 2(\lambda/2\ell_1)^2, \quad \alpha_2 = C(\lambda/2\ell_1)^2, \quad 0 < C < 2.$$

Proof of Theorem 2.1. It is easy to verify the following chain of relations:

$$P\{\sup_{a\le x\le b} |r_n(x) - r(x)| > \lambda\} \le P\{\sup_{a\le x\le b} |\phi_n(x) - r(x)g_n(x)| \ge$$

$$\ge \lambda(\mu-\lambda)\} + P\{\sup_{a\le x\le b} |g_n(x) - g(x)| > \lambda\} \le P\{\sup_{a\le x\le b} |\phi_n(x) -$$

$$- r(x)g(x) + \sup_{a\le x\le b} |r(x)g_n(x) - r(x)g(x)| \ge \lambda(\mu-\lambda)\} + P\{\sup_{a\le x\le b} \times$$

$$\times |g_n(x) - g(x)| > \lambda\} \le P\left\{\sup_{a\le x\le b} |\phi_n(x) - \phi(x)| \ge \frac{\lambda(\mu-\lambda)}{2}\right\} + P\left\{\sup_{a\le x\le b} \times\right.$$

$$\left.\times |g_n(x) - g(x)| > \frac{\lambda(\mu-\lambda)}{2\ell_2}\right\} + P\{\sup_{a\le x\le b} |g_n(x) - g(x)| > \lambda\}, \quad (2.4)$$

where $\ell_2 = \max_{a\le x\le b} |r(x)|$. Utilizing Lemmas 2.1 and 2.2 one easily obtains

$$P\{\sup_{a\le x\le b} |\phi_n(x) - \phi(x)| > \lambda\} \le P\{\sup_{a\le x\le b} |\phi_n(x) - E\phi_n(x) > \lambda/2\} \le$$

$$\le C_0 e^{-\beta_1 n a_n^{-2}} + C_1 e^{-\beta_2 n a_n^{-2}} \quad \text{for} \quad n > N_1(\lambda), \quad (2.5)$$

where $\beta_1 = 2(\lambda/4\ell_1)^2$ and $\beta_2 = C(\lambda/4\ell_1)^2$. On the other hand, utilizing inequalities (2.1.5) and (2.1.6) it is easy to establish the inequality

$$P\{\sup_{a\le x\le b} |g_n(x) - g(x)| > \lambda\} \le C_0 e^{-a_1 n a_n^{-2}} \quad \text{for} \quad n > N_2(\lambda). \quad (2.6)$$

Finally in view of (2.5) and (2.6) we obtain from (2.4) that

$$P\{\sup_{a\le x\le b} |r_n(x) - r(x)| > \lambda\} \le C_0(e^{-\gamma_1 n a_n^{-2}} + e^{-\gamma_2 n a_n^{-2}} + e^{-\gamma_3 n a_n^{-2}}) +$$

$$+ C_1 e^{-\gamma_4 n a_n^{-2}}$$

for

$$n > N(\lambda) = \max(N_1(\lambda), N_2(\lambda)), \quad (2.7)$$

where

$$\gamma_1 = 2(\lambda(\mu-\lambda)/8\ell_1)^2, \quad \gamma_2 = 2(\lambda(\mu-\lambda)/4\ell_1\ell_2)^2,$$

$$\gamma_3 = 2(\lambda/2\ell_1)^2, \quad \gamma_4 = C(\lambda(\mu-\lambda)/8\ell_1)^2.$$

Inequality (2.7) with the aid of the Borel-Cantelli lemma completes the proof of the theorem.

3. Limiting Distribution of the Maximal Deviation of Estimators of Regression Curves

1. In this section limit theorems for the maximal deviation of nonparametric estimators $r_n(x)$ of a regression curve (as defined in Section 1 of Chapter 4) from the theoretical curve $r(x)$ are proved. These theorems allow us to construct for the unknown regression curve $r(x)$ a class of confidence regions with a given confidence coefficient and also to test the null hypothesis H_0 that the regression function $r(x)$ is equal to a function $r_0(x)$ given in advance.

In paper [80] we have solved the problem of deriving the limiting distribution of the maximal deviation of the estimator $r_n(x)$ from $r(x)$ on a finite interval using the same general method of proof as the one presented in Section 1 of Chapter 3. In [24] using approximately the same logical method of proof, Lemma 3.3 and Theorem 3.3 presented below were extended to the multivariate case.

2. Throughout this section the following assumptions are made:

1°. $f(x, y) = 0$ for $|y| > A$, where A is a constant, i.e. the random variable Y is bounded a.s. The function $g(x) = \int f(x, y)\,dy$ is bounded on the whole axis, $g(x)$, $\phi(x) = \int yf(x, y)\,dy$ and $V(x) = \int y^2 f(x, y)\,dy$ are continuous on the interval $-\infty < a \leq x \leq b < \infty$;

2°.
$$\min_{a\leq x\leq b} g(x) > 0, \qquad \min_{a\leq x\leq b} \beta(x) > 0, \qquad \text{where} \quad \beta(x) = D(Y/X = x);$$

3°. the kernel $K(x)$ is a density function which satisfies the following conditions: it is of a bounded variation, $K(-x) = K(x)$, $K(|x_2|) \leq K(|x_1|)$ for $|x_2| > |x_1|$ and $\int |u|K(u)\,du < \infty$.

Assertion a) of Theorem 4.1.4 motivates the transition from the difference $r_n(x) - r(x)$ to the sum $\eta_n(x) = 1/n \sum_{i=1}^{n} \eta_n^{(i)}(x)$ of a sequence of series of independent random variables

$$\eta_n^{(i)}(x) = a_n(Y_i - r(x))K(a_n(x - X_i)), \qquad i = 1, 2, \ldots$$

LEMMA 3.1. Uniformly in x, $a \leq x \leq b$,

$$a_n^{-1} D\eta_n^{(i)}(x) \to \beta(x)g(x) \int K^2(u)\,du \quad \text{as} \quad n \to \infty. \tag{3.1}$$

<u>Proof</u>. Since, in view of Lemma 4.2.1 $E\eta_n^{(i)}(x) = O(1)$ uniformly in x it follows that

$$\begin{aligned} a_n^{-1} D\eta_n^{(i)}(x) &= a_n \iint (y - r(x))^2 K^2(a_n(x - u)) f(u, y)\,du\,dy + \\ &+ O(a_n^{-1}) = a_n \int K^2(a_n(x - u))V(u)\,du - 2r(x)a_n \times \\ &\times \int K^2(a_n(x - u))\phi(u)\,du + r^2(x)a_n \int K^2(a_n(x - \\ &- u))g(u)\,du + O(a_n^{-1}). \end{aligned} \tag{3.2}$$

Analogously one obtains from the same lemma that the integrals on the r.h.s. of (3.2) converge uniformly in x, $a \le x \le b$, to the quantities $V(x) \int K^2(u)\,du$, $\phi(x) \int K^2(u)\,du$ and $g(x) \int K^2(u)\,du$, respectively. Hence

$$D\eta_n^{(i)}(x) \sim a_n(V(x)/g(x) - r^2(x))g(x) \int K^2(u)\,du =$$

$$= a_n\beta(x)g(x) \int K^2(u)\,du,$$

q.e.d.

We subdivide the interval $a \le x \le b$ into s_n subintervals $\Delta_1, \Delta_2, \ldots, \Delta_{s_n}$ of the same length τ_n. Below the number of subintervals s_n will increase as the sample size n increases.

LEMMA 3.2. Let t_i, $i = \overline{1, s_n}$ be the centers of intervals Δ_i, $i = \overline{1, s_n}$ and

$$\zeta_n(t_j) = \frac{\eta_n(t_j) - E\eta_n(t_j)}{\sqrt{D\eta_n(t_j)}} = n^{-1/2} \sum_{i=1}^{n} \xi_n^{(i)}(t_j), \quad 1 \le j \le s_n,$$

where

$$\xi_n^{(i)}(t_j) = a_n^{-1/2}[\eta_n^{(i)}(t_j) - E\eta_n^{(i)}(t_j)]\, d_n^{-1}(t_j),$$

$$d_n(t_j) = (a_n^{-1} D\eta_n^{(i)}(t_j))^{1/2}.$$

Then

$$\varepsilon_n = \max_{1 \le i,j \le s_n, i \ne j} |E(\zeta_n(t_j)\zeta_n(t_i))| \le C_1 K^*(\tau_n a_n) + C_2 a_n^{-1}, \quad (3.3)$$

where

$$K^*(x) = K(x) * K(x).$$

Proof. We have

$$E(\zeta_n(t_i)\zeta_n(t_j)) = a_n^{-1}(E\eta_n^{(1)}(t_j)\eta_n^{(1)}(t_i) - E\eta_n^{(1)}(t_i)E\eta_n^{(1)}(t_j)) \times$$

$$\times d_n^{-1}(t_j)\, d_n^{-1}(t_i), \quad (3.4)$$

and moreover $a_n^{-1}|E\eta_n^{(1)}(t_j)E\eta_n^{(1)}(t_i)| \le C_3 a_n^{-1}$ uniformly in t_j. Taking condition 1° into account we obtain

$$a_n^{-1}|E(\eta_n^{(1)}(t_j)\eta_n^{(1)}(t_i))| \le C_4 \int K(u)K(a_n|t_i - t_j| - u)\,du =$$

$$= C_4 K^*(a_n(|t_i - t_j|)). \quad (3.5)$$

Property 3° of function K(x) implies that $K^*(|x_1|) \ge K^*(|x_2|)$ for $|x_1| < |x_2|$. Therefore we infer from (3.5) that

$$E|\eta_n^{(1)}(t_j)\eta_n^{(1)}(t_i)| \leq C_4 K^*(a_n\tau_n), \tag{3.6}$$

since

$$\min_{i\neq j} |t_i - t_j| \geq \tau_n.$$

Next, from Lemma 3.1 and condition 2° we obtain

$$d_n(t_j) \geq C_0 > 0 \quad \text{for all} \quad t_j,\ j = \overline{1, s_n}. \tag{3.7}$$

Thus relations (3.4)-(3.7) prove our lemma.

Directly from conditions 2° and 3° follows

LEMMA 3.3. $\max\limits_{1\leq i\leq n} \sup\limits_{a\leq x\leq b} |\xi_n^{(i)}(x)| \leq C_5 a_n^{1/2}$ a.s.

The next lemma whose proof is based on Lemmas 3.2 and 3.3 establishes the behavior of probabilities of large deviations of the sum $\eta_n(x)$.

LEMMA 3.4. Let λ_{s_n} be a nonnegative and increasing with s_n function of s_n; r be a fixed integer less than s_n; $k_1, k_2, \ldots, k_r$ be any combination of r numbers from the sequence 1, 2, ..., s_n, $m = \max(3, r)$.

If

$$\varepsilon_n\lambda_{s_n}^2 \to 0, \quad \lambda_n^m/\sqrt{na_n^{-1}} \to 0 \quad \text{as} \quad n \to \infty,$$

we then have

$$P\{|\zeta_n(t_j)| \geq \lambda_{s_n}, \quad j = 1, 2, \ldots, r\} = \left(\frac{2}{\sqrt{2\pi}}\int_{\lambda_{s_n}}^{\infty} e^{-u^2/2}\,du\right)^r \times$$
$$\times \left[1 + O(\varepsilon_n\lambda_{s_n}^2) + O\left(\frac{\lambda_{s_n}^m}{\sqrt{na_{s_n}^{-1}}}\right)\right].$$

The proof of this lemma is similar to the proof of Lemma 3.1.3 and is therefore omitted.

Using the method of proving Theorem 3.1.5 one can easily prove with the aid of Lemmas 3.1.4, 3.1.5 and 3.4 the following

THEOREM 3.1. If $\varepsilon_n \log s_n \to 0$ and $(\log s_n)^m/na_n^{-1} \to 0$, $m = 0, 1, 2, \ldots$ as $n \to \infty$, then

$$P\left\{\max_{1\leq i\leq s_n}\left|\frac{\eta_n(t_i) - E\eta_n(t_i)}{(D\eta_n(t_i))^{1/2}}\right| < \lambda_{s_n}\right\} \to e^{-2e^{-\lambda}},$$

where $\lambda_{s_n} = \ell_{s_n} + \lambda/\ell_{s_n}$ and ℓ_{s_n} is the root of the equation

$$1/s_n = (2\pi)^{-1/2} \int_{\ell_{s_n}}^{\infty} \exp\left\{-\frac{x^2}{2}\right\} dx, \quad (\ell_{s_n} = O\sqrt{\log s_n})).$$

Theorem 3.1 yields

THEOREM 3.2. Let $g(x)$, $\phi(x)$ and $V(x)$ possess bounded first order derivatives. If $a_n^{-3} n \log s_n \to 0$, $\varepsilon_n \log s_n \to 0$ and $n^{-1} a_n (\log s_n)^m \to 0$ as $n \to \infty$, $m = 0, 1, 2, \ldots$, then

$$P\left\{\sqrt{\frac{n}{a_n}} \max_{1 \le i \le s_n} \left| \frac{\eta_n(t_i)}{\sqrt{\beta(t_i) g(t_i)}} \right| < \left(\ell_{s_n} + \frac{\lambda}{\ell_{s_n}}\right) \int K^2(u)\, du\right\} \to e^{-2e^{-\lambda}}$$

To prove this theorem it is sufficient to note that

$$E\eta_n(t) = O(a_n^{-1}) \quad \text{and} \quad a_n^{-1} D\eta_n^{(1)}(t) = \beta(t) g(t) \int K^2(u)\, du + O(a_n^{-1})$$

uniformly in t, $a \le t \le b$.

Remark to Theorem 3.2. If one assumes that $g(x)$ and $\phi(x)$ possess bounded second-order derivatives and $x^2 K(x) \in L_1(-\infty, \infty)$ then the condition $a_n^{-3} n \log s_n \to 0$ is replaced by a weaker condition $a_n^{-5} n \log s_n \to 0$ which will appear in Theorem 3.7.

THEOREM 3.3. Under the conditions of Theorem 3.2

$$\lim_{n\to\infty} P\left\{\sqrt{\frac{n}{a_n}} \max_i \left| \frac{r_n(t_i) - r(t_i)}{\sigma(t_i)} \right| < (\ell_{s_n} + \lambda/\ell_{s_n}) K_0 \right\} = e^{-2e^{-\lambda}},$$

where

$$\sigma^2(x) = \beta(x)/g(x) \quad \text{and} \quad K_0^2 = \int K^2(x)\, dx.$$

Proof. Let $T_n(x) = g_n(t_i)$ for $x \in \Delta_i$, $i = \overline{1, s_n}$ and set

$$z_n = \max_{a \le x \le b} |(T_n(x) - g(x)) g^{-1/2}(x)|.$$

Then taking condition 2° into account it is easy to verify that

$$\ell_{s_n}^2 \max_i |g_n(t_i) g^{-1}(t_i) - 1| \le \mu \max_i |(g_n(t_i) - g(t_i)) g^{-1/2} \times$$

$$\times (t_i)| \le \mu [A_n(\ell_{s_n} \sqrt{\frac{n}{a_n}} z_n - \ell_{s_n}^2) + B_n], \tag{3.8}$$

where

$$A_n = \ell_{s_n} (a_n/n)^{1/2}, \quad B_n = \ell_{s_n}^2 (a_n/n)^{1/2} \quad \text{and } \mu = \min_{a \le x \le b} g(x).$$

However, in view of Theorem 3.1.6, $\ell_{s_n} \sqrt{n/a_n}\, z_n - \ell_{s_n}^2$ possesses an asymptotic distribution and by the condition of the theorem $A_n \to 0$ and $B_n \to 0$ as $n \to \infty$. Hence these facts together with (3.8) imply that

$$\ell_{s_n}^2 \max_i |g_n(t_i) g^{-1}(t_i) - 1| \to 0 \tag{3.9}$$

in probability as $n \to \infty$.

Denote

$$U_n = \sqrt{\frac{n}{a_n}} \max_i \left| \frac{\eta_n(t_i)}{\sqrt{\beta(t_i) g(t_i)}} \right|,$$

$$M_n = \sqrt{\frac{n}{a_n}} \max_i \left| \frac{r_n(t_i) - r(t_i)}{\sigma(t_i)} \right|,$$

$$h_n = \min_i (g_n(t_i)/g(t_i) - 1),$$

$$H_n = \max_i (g_n(t_i)/g(t_i) - 1).$$

Noting that $\eta_n(t) = \phi_n(t) - r(t) g_n(t)$ and using the identity

$$\frac{r_n(x) - r(x)}{\sigma(x)} = \frac{\eta_n(x)}{\sqrt{\beta(x) g(x)}} \; \frac{1}{1 + (g_n(x)/g(x) - 1)},$$

it is easy to derive the inequalities

$$P\{U_n \le \lambda_{s_n}(1 + h_n)\} \le P\{M_n \le \lambda_{s_n}\} \le P\{U_n \le \lambda_{s_n}(1 + H_n)\}, \tag{3.10}$$

where

$$\lambda_{s_n} = \lambda_{s_n} + \lambda/\ell_{s_n}.$$

Moreover

$$P\{U_n \le \lambda_{s_n}(1 + h_n)\} = P\{(\ell_{s_n} U_n - \ell_{s_n}^2) z_n + \theta_n < \lambda\},$$

where

$$z_n = 1/(1 + h_n), \quad \theta_n = -h_n \ell_{s_n}^2 / (1 + h_n).$$

In view of (3.9) z_n converges in probability to one and θ_n converges to zero as $n \to \infty$. Using these facts and applying Theorem 3.2 we obtain the following result

$$\lim_{n\to\infty} P\{U_n < \lambda_{s_n}(1 + h_n)\} = e^{-2e^{-\lambda}} \tag{3.11}$$

It can be shown analogously that

$$\lim_{n\to\infty} P\{U_n < \lambda_{s_n}(1 + H_n)\} = e^{-2e^{-\lambda}}. \tag{3.12}$$

Thus (3.10), (3.11) and (3.12) complete the proof of the theorem.

3. Define the function $m_n(x)$, $a \leq x \leq b$, as follows:

$$m_n(x) = r_n(t_i), \quad x \in \Delta_i, \quad i = \overline{1, s_n}.$$

THEOREM 3.4. Let $g(x)$, $\phi(x)$ and $V(x)$ satisfy the conditions of Theorem 3.2. Moreover, let the following condition be fulfilled as $n \to \infty$: $\varepsilon_n \log s_n \to 0$, $a_n^{-3} n \log s_n \to 0$, $n^{-1} a_n (\log s_n)^m \to 0$, for any fixed m, m = 0, 1, 2, ..., $(n \log s_n)/a_n s_n^2 \to 0$. Then

$$\lim_{n\to\infty} P\left\{ \max_{a\leq x\leq b} \left| \frac{m_n(x) - r(x)}{\sigma(x)} \right| < \left(\ell_{s_n} + \frac{\lambda}{\ell_{s_n}} \right) K_0 \sqrt{\frac{a_n}{n}} \right\} = e^{-2e^{-\lambda}}. \tag{3.13}$$

Proof. The conditions of the theorem imply that

$$r(x) - r(t_i) = O(s_n^{-1}) \quad \text{and} \quad \sigma(x) - \sigma(t_i) = O(s_n^{-1}) \quad \text{for} \quad x \in \Delta_i \tag{3.14}$$

Set

$$M_n^* = \max_{a\leq x\leq b} \left| \frac{m_n(x) - r(x)}{\sigma(x)} \right|, \quad \delta_n = M_n^* - M_n$$

and show that $\ell_{s_n} \sqrt{n/a_n}\, |\delta_n| \to 0$ in probability. Indeed, we have

$$\left| \left| \frac{m_n(x) - r(x)}{\sigma(x)} \right| - \left| \frac{r_n(t_i) - r(t_i)}{\sigma(t_i)} \right| \right| \leq \left| \frac{r(x) - r(t_i)}{\sigma(s_i)} \right| +$$

$$+ \left| \frac{m_n(x) - r(x)}{\sigma(t_i)} - \frac{m_n(x) - r(x)}{\sigma(x)} \right| \leq \left| \frac{r(x) - r(t_i)}{\sigma(t_i)} \right| +$$

$$+ |m_n(x) - r(x)| \left| \frac{\sigma(x) - \sigma(t_i)}{\sigma(t_i)\sigma(x)} \right| \quad \text{for} \quad x \in \Delta_i. \tag{3.15}$$

Since in view of condition 1° there exists a constant C_6 such that $|Y| \leq C_6$ a.s. it follows that $|r_n(x)| \leq C$ a.s. for all x and n. Consequently, for $x \in \Delta_i$

$$|m_n(x) - r(x)| \leq |r_n(t_i) - r(t_i)| + |r(t_i) - r(x)| \leq C_7 + C_8 a_n^{-1} \leq C_9$$

a.s.

This together with (3.14)-(3.15) implies that $|\delta_n| \le C_{10} s_n^{-1}$ a.s. On the other hand, from the condition

$$n \log s_n / a_n s_n^2 \to 0, \quad \ell_{s_n} \sqrt{\frac{n}{a_n}}\, |\delta_n| \le C_{11} \sqrt{\frac{n \log s_n}{a_n s_n^2}} \to 0$$

as $n \to \infty$. Therefore for any $\varepsilon > 0$

$$P\{\ell_{s_n} \sqrt{\frac{n}{a_n}}\, |\delta_n| > \varepsilon\} \to 0 \quad \text{as} \quad n \to \infty .$$

Taking inequalities

$$P\left\{\sqrt{\frac{n}{a_n}}\, M_n^* \le \left(\ell_{s_n} + \frac{\lambda}{\ell_{s_n}}\right) K_0\right\} \ge P\left\{\sqrt{\frac{n}{a_n}}\, M_n < \left(\ell_{s_n} + \frac{\lambda - \varepsilon}{\ell_{s_n}}\right) \times \right.$$

$$\left. \times K_0\right\} - P\left\{\sqrt{\frac{n}{a_n}}\, |\delta_n| > \frac{\varepsilon}{\ell_{s_n}} K_0\right\} \tag{3.16}$$

and

$$P\left\{\sqrt{\frac{n}{a_n}}\, M_n^* \le \left(\ell_{s_n} + \frac{\lambda}{\ell_{s_n}}\right) K_0\right\} \le P\left\{\sqrt{\frac{n}{a_n}}\, M_n < \left(\ell_{s_n} + \frac{\lambda + \varepsilon}{\ell_{s_n}}\right) \times \right.$$

$$\left. \times K_0\right\} + P\left\{\sqrt{\frac{n}{a_n}}\, |\delta_n| > \frac{\varepsilon}{\ell_{s_n}} K_0\right\},$$

and the result of Theorem 3.3 into account we arrive at

$$e^{-2e^{-(\lambda-\varepsilon)}} \le \varliminf_{n\to\infty} P\left\{\sqrt{\frac{n}{a_n}}\, M_n^* \le \left(\ell_{s_n} + \frac{\lambda}{\ell_{s_n}}\right) K_0\right\} \le \varlimsup_{n\to\infty} P \times$$

$$\times \left\{\sqrt{\frac{n}{a_n}}\, M_n^* \le \left(\ell_{s_n} + \frac{\lambda}{\ell_{s_n}}\right) K_0\right\} \le e^{-2e^{-(\lambda+\varepsilon)}} .$$

Where since $\varepsilon > 0$ is arbitrary we obtain

$$\lim_{n\to\infty} P\left\{\sqrt{\frac{n}{a_n}}\, M_n^* \le \left(\ell_{s_n} + \frac{\lambda}{\ell_{s_n}}\right) K_0\right\} = e^{-2e^{-\lambda}} .$$

q.e.d.

THEOREM 3.5. Under the conditions of Theorem 3.4

$$\lim_{n\to\infty} P\left\{\sqrt{\frac{n}{a_n}} \max_{a\le x\le b} \left|\frac{m_n(x) - r(x)}{\overline{\sigma_n}(x)}\right| < (\ell_{s_n} + \lambda/\ell_{s_n}) K_0\right\} = e^{-2e^{-\lambda}}, \tag{3.17}$$

where $\overline{\sigma_n}(x) = \sigma(t_i)$ for $x \in \Delta_i$, $i = 1, 2, \ldots, s_n$.

The proof of this theorem is completely analogous to the proof of Theorem 3.4 and is therefore omitted.

We now explain how to select a_n and s_n so that the conditions of Theorem 3.4 will be fulfilled. For this purpose two cases will be considered.

Case 1. Let the function $K(x)$ be finite, i.e. $K(x) = 0$ for $|x| > x_0$ and $s_n = (b - a)/C_0 x_0 \cdot a_n$, $C_0 \geq 2$. Then $\tau_n = C_0 x_0 a_n^{-1}$ and $K^*(\tau_n \cdot a_n) = K^*(C_0 \cdot x_0) = 0$. Hence $\varepsilon_n = O(a_n^{-1})$ and thus the condition $\varepsilon_n \log s_n \to 0$ of Theorem 3.4 will be fulfilled if one sets $a_n = a_0 n^{\alpha}$, where $1/3 < \alpha < 1$.

As an example we shall consider the kernel $K(x)$ of the form: $K(x) = 1/2 (= 0)$ for $|x| \leq 1$ ($|x| > 1$), respectively. Then

$$m_n(x) = r_n(t_j) = \frac{1}{N_j} \sum_{x_i \in \Delta_j} Y_i \quad \text{for} \quad x \in \Delta_j, \tag{3.18}$$

where N_i is the number of X's falling into Δ_i. Evidently (3.18) is an analog of a 'histogram'.

Case 2. Let $K(x)$ satisfy only the conditions 3° and $\tau_n \cdot a_n \to \infty$, i.e. $s_n^{-1} a_n \to \infty$ as $n \to \infty$. It follows from the properties of $K(x)$ that $xK^*(x) \to 0$ as $|x| \to \infty$. Then

$$\varepsilon_n \leq C_1(\tau_n \cdot a_n) K^*(\tau_n a_n) \frac{1}{\tau_n a_n} + C_2 \frac{1}{\tau_n a_n} \tau_n = O\left(\frac{1}{\tau_n \cdot a_n}\right). \tag{3.19}$$

Therefore the conditions of Theorem 3.4 will be fulfilled if we set $a_n = a_0 n^{\alpha}$, $s_n = s_0 n^{\beta}$ where $1/3 < \alpha < 1$, $(1 - \alpha)/2 < \beta < \alpha$.

In Theorem 3.4 we have considered the maximum of normalized deviations of the regression curve $r(x)$ from $m_n(x)$. However, in place of $m_n(x)$ one could consider a more accurate approximating continuous function $\tilde{m}_n(x)$ defined as follows:

$$\tilde{m}_n(x) = r_n(t_k) + \tau_n^{-1}(x - t_k)(r_n(t_{k+1}) - r_n(t_k)) \quad \text{as} \quad x \in \Delta_k.$$

If $a_0 = a, a_1, \ldots, a_{s_n} = b$ are the end points of the intervals Δ_i, $i = \overline{1, s_n}$, one could then write $\tilde{m}_n(x)$ in the form

$$\tilde{m}_n(x) = 1/2(r_n(a_k + \tau_n/2) + r_n(a_k - \tau_n/2)) + \tau_n^{-1}(x - a_k) \times$$
$$\times [r_n(a_k + \tau_n/2) - r_n(a_k - \tau_n/2)].$$

In an analogous manner we introduce

$$\tilde{m}(x) = 1/2(r(a_k + \tau_n/2) + r(a_k - \tau_n/2)) + \tau_n^{-1}(x - a_k) \times$$
$$\times [r(a_k + \tau_n/2) - r(a_k - \tau_n/2)],$$

$$\tilde{\sigma}(x) = 1/2(\sigma(a_k + \tau_n/2) + \sigma(a_k - \tau_n/2)) + \tau_n^{-1}(x - a_k) \times$$
$$\times [\sigma(a_k + \tau_n/2 - \sigma(a_k - \tau_n/2)].$$

Consider now

$$\tilde{M}_n = \max \left| \frac{\tilde{m}_n(x) - \tilde{m}(x)}{\tilde{\sigma}(x)} \right|, \quad a + \tau_n/2 \leq x \leq b - \tau_n/2.$$

It is easy to verify that this maximum is attained at one of the end-points of the segment $[t_k, t_{k+1}]$ (cf. Subsection 7 in Section 1 of Chapter 3). Hence $M_n = \tilde{M}_n$. Noting this fact and taking the result of Theorem 3.3 into account we can state the following theorem.

THEOREM 3.6. Under the conditions of Theorem 3.3

$$\lim_{n\to\infty} P\{\tilde{M}_n < (\ell_{s_n} + \lambda/\ell_{s_n})K_0\} = e^{-2e^{-\lambda}}.$$

THEOREM 3.7. Let $g(x)$, $\phi(x)$ and $V(x)$ possess second order bounded derivatives. Moreover, let $x^2K(x) \in L_1$ and as $n \to \infty$ the following relations will be fulfilled: $\varepsilon_n \log s_n \to 0$,

$$a_n^{-5} n \log s_n \to 0, \quad n \log s_n/a_n s_n^4 \to 0 \text{ and } n^{-1}a_n(\log s_n)^m \to 0$$

for any fixed m, $m = 0, 1, 2, \ldots$. Then

$$\lim_{n\to\infty} P\left\{\sqrt{\frac{n}{a_n}} \max_{a+\tau_n/2\leq x\leq b-\tau_n/2} \left|\frac{\tilde{m}_n(x) - r(x)}{\sigma(x)}\right| \leq \left(\ell_{s_n} + \frac{\lambda}{\ell_{s_n}}\right) K_0 \right\} =$$
$$= e^{-2e^{-\lambda}}. \tag{3.20}$$

Proof. Denote

$$\tilde{\tilde{M}}_n = \max \left| \frac{\tilde{m}_n(x) - r(x)}{\sigma(x)} \right|, \quad (a + \tau_n/2 \leq x \leq b - \tau_n/2),$$

$$\delta_n = \tilde{\tilde{M}}_n - \tilde{M}_n.$$

From the condition of the theorem one can write

$$r(a_k \pm \tau_n/2) = r(a_k) \pm \frac{\tau_n}{2} r'(a_k) + O\left(\frac{1}{s_n}\right)$$

and at each subinterval Δ_i

$$\tilde{m}(x) = r(a_k) + (x - a_k)r'(a_k) + O\left(\frac{1}{s_n^2}\right),$$

$$r(x) = r(a_k) + (x - a_k)r'(a_k) + O\left(\frac{1}{s_n^2}\right).$$

Hence

$$r(x) - \tilde{m}(x) = O(s_n^{-2}). \tag{3.21}$$

In the same manner one can show that

$$\sigma(x) - \tilde{\sigma}(x) = O(s_n^{-2}). \tag{3.22}$$

In view of (3.21), (3.22) and an inequality analogous to (3.15) we deduce that $|\delta_n| \leq C_{12}s_n^{-2}$ a.s. Hence

$$\ell_{s_n}\sqrt{\frac{n}{a_n}}\,\delta_n \leq C_{13}\left(\frac{n \log s_n}{a_n \cdot s_n^4}\right)^{1/2} \quad \text{a.s.}$$

Thus $\ell_{s_n}\sqrt{n/a_n}\,\delta_n \to 0$ in probability as $n \to \infty$. To complete the proof it remains only to replace M_n by $\tilde{M}_n$ and M_n^* by $\tilde{\tilde{M}}_n$ in the inequalities (3.16) and then apply the result of Theorem 3.5. The theorem is thus proved.

The conditions of Theorem 3.7 in regard to a_n and s_n - in view of Theorem (3.19) - are fulfilled, for example, if one sets $a_n = a_0 n^{\alpha}$ and $s_n = s_0 n^{\beta}$ for $1/5 < \alpha < 1$, $(1 - \alpha)/4 < \beta < \alpha$.

4. We shall consider statistical applications of the results obtained for solving the two basic problems: I) construction of the asymptotic confidence region for the regression curve $r(x)$; II) construction of tests for testing the null hypothesis H_0 that $r(x) = r_0(x)$ ($r_0(x)$ is a given function).

I. In the first section of this chapter we were dealing with the construction of confidence intervals for $r(x)$ corresponding to a specific value of x. Assume now that it is required to construct a confidence region for the whole graph of the regression curve on the interval $a \leq x \leq b$ (a and b are given numbers), i.e. a planar region G such that with probability α, $0 < \alpha < 1$, the true regression curve $r(x)$ is contained in G. This problem is quite different from the one mentioned above.

The problem posed - in the simplest parametric case i.e. the case of a linear regression model corresponding to the normal distribution - was solved in [57]. In our nonparametric case it is solved with the aid of Theorems 3.4 and 3.7. We shall consider several cases.

a_1. Confidence region for $r(x)$ when $\sigma(x)$ is known. In this case it follows from (3.13) that with probability arbitrarily close to α, the region $G_\alpha(K)$ is bounded by the curves $h_n^{\pm}(x) = m_n(x) \pm \lambda'_\alpha \sigma(x)$ for all x on the interval $a \le x \le b$, where $\lambda'_\alpha = (\ell_{s_n} + \lambda_\alpha/\ell_{s_n})K_0$ and λ_α is the solution of the equation $\exp(-2\exp - \lambda) = \alpha$ may serve as a confidence region for $r(x)$. In exactly the same manner one determines from (3.20) the region $\tilde{G}_\alpha(K)$ and the curves

$$\tilde{h}_n^{\pm}(x) = \tilde{m}_n(x) \pm \lambda_\alpha \sigma(x).$$

a_2. Confidence region for $r(x)$ when $\sigma(x)$ is unknown. In this case $\sigma(x)$ must be estimated. As we have seen the quantity

$$\sigma_n^2(x) = (q_n(x) - r_n^2(x))g_n^{-1}(x)$$

(cf. (4.1.19)) where

$$q_n(x) = \sum_{i=1}^{n} Y_i^2 K(a_n(x - X_i)) \left[\sum_{i=1}^{n} K(a_n(x - X_i))\right]^{-1},$$

is a consistent estimator of $\sigma^2(x)$.

THEOREM 3.8. Let $g(x)$, $\phi(x)$ and $V(x)$ satisfy the conditions of Theorem 3.4. If in addition to the conditions of Theorem 3.3 with regard to a_n, s_n and n one also requires that

$$\lim_{n\to\infty} P\left\{\sqrt{\frac{n}{a_n}} \sup_{a\le x\le b} \left|\frac{m_n(x) - r(x)}{\sigma_n(x)}\right| < \left(\ell_{s_n} + \frac{\lambda}{\ell_{s_n}}\right)K_0\right\} = \exp(-2e^{-\lambda}), \tag{3.23}$$

then

$$\lim_{n\to\infty} P\left\{\sqrt{\frac{n}{a_n}} \sup_{a\le x\le b} \left|\frac{m_n(x) - r(x)}{\sqrt{q_n^2(x) - r^2(x)}}\right| < \left(\ell_{s_n} + \frac{\lambda}{\ell_{s_n}}\right)K_0\right\} =$$

$$= \exp(-2e^{-\lambda}). \tag{3.24}$$

Proof. The proof is based on the same ideas which were utilized in Theorem 3.4.

Introduce the notation

$$\beta_n(x) = q_n(x) - r_n^2(x), \qquad V_n(x) = \frac{a_n}{n}\sum_{i=1}^{n} Y_i^2 K(a_n(x - X_i)).$$

It follows from inequalities (4.2.6) and (4.2.7) that respectively

$$\log^2 s_n \max_{a\le x\le b} |g_n(x) - g(x)| \to 0 \text{ and } \log^2 s_n \max_{a\le x\le b} |r_n(x) - r(x)| \to 0$$

in probability as $n \to \infty$. A similar inequality is also valid for

$\max_{a \leq x \leq b} |V_n(x) - V(x)|$. Consequently $\log^2 s_n \max_{a \leq x \leq b} |V_n(x) - V(x)| \to 0$ in probability as $n \to \infty$. From here and the inequality

$$\sup_{a \leq x \leq b} |\beta_n(x) - \beta(x)| = \sup_{a \leq x \leq b} |V_n(x) g_n^{-1}(x) - V(x) g^{-1}(x) + r_n^2(x) -$$
$$- r^2(x)| \leq \sup_{a \leq x \leq b} |(V_n(x) - V(x)) g^{-1}(x)| + C_{14} \sup_{a \leq x \leq b} \times$$
$$\times |(g_n(x) - g(x)) g^{-1}(x)| + C_{15} \sup_{a \leq x \leq b} |r_n(x) - r(x)| \leq$$
$$\leq C_{16} \sup_{a \leq x \leq b} |V_n(x) - V(x)| + C_{17} \sup_{a \leq x \leq b} |g_n(x) - g(x)| +$$
$$+ C_{18} \sup_{a \leq x \leq b} |r_n(x) - r(x)|,$$

which follows from conditions 1° and 2° we have

$$\lim_{n \to \infty} \log^2 s_n \sup_{a \leq x \leq b} |\beta_n(x) - \beta(x)| = 0 \tag{3.25}$$

in probability.

Next for n large

$$\sup_{a \leq x \leq b} \left| \frac{\sigma_n^2(x)}{\sigma^2(x)} - 1 \right| \leq$$
$$\leq \frac{C_{19} \sup_{a \leq x \leq b} |\beta_n(x) - \beta(x)| + C_{20} \sup_{a \leq x \leq b} |g_n(x) - g(x)|}{\mu - \sup_{a \leq x \leq b} |g_n(x) - g(x)|},$$

where $\mu = \min_{a \leq x \leq b} g(x)$. Whence taking (3.25) into account we obtain that

$$\lim_{n \to \infty} \log^2 s_n \sup_{a \leq x \leq b} \left| 1 - \frac{\sigma_n^2(x)}{\sigma^2(x)} \right| = 0 \tag{3.26}$$

in probability.

For any real numbers $\alpha, \beta > 0$

$$|\sqrt{\alpha} - \sqrt{\beta}|^2 \leq |\alpha - \beta|.$$

Therefore in view of (3.26)

$$\lim_{n \to \infty} \log s_n \sup_{a \leq x \leq b} \left| \sqrt{\frac{\sigma_n^2(x)}{\sigma^2(x)}} - 1 \right| = 0 \tag{3.27}$$

in probability.

Denote

$$\overline{M}_n = \sqrt{\frac{n}{a_n}} \sup_{a \le x \le b} \left| \frac{m_n(x) - r(x)}{\sigma_n(x)} \right| ,$$

$$\overline{h}_n = \inf_{a \le x \le b} \left(\frac{\sigma_n(x)}{\sigma(x)} - 1 \right) ,$$

$$\overline{H}_n = \sup_{a \le x \le b} \left(\frac{\sigma_n(x)}{\sigma(x)} - 1 \right) .$$

Then

$$P\{M_n \le \lambda_{s_n}(1 + \overline{h}_n)\} \le P\{\overline{M}_n \le \lambda_{s_n}\} \le P\{M_n \le \lambda_{s_n}(1 + \overline{H}_n)\}, \quad (3.28)$$

where $\lambda_{s_n} = \ell_{s_n} + \lambda/\ell_{s_n}$.

Analogously to relations (3.11) and (3.12) utilizing (3.27) it can be shown that the limits as $n \to \infty$ of the first and the third probabilities appearing in the inequality (3.28) are the same and are equal to $\exp(-2e^{-\lambda})$. Thus (3.23) is proved. Assertion (3.24) is verified using verbatim the same arguments as those presented in the proof of (3.23).

It follows from (3.23) that with probability as close to α as desired and for n sufficiently large the region $\tilde{G}_\alpha(K)$ bounded by the curves

$$\tilde{h}_n^{\pm}(x) = m_n(x) \pm \lambda'_\alpha \sigma_n(x), \quad \text{where } \lambda'_\alpha = (\ell_{s_n} + \lambda_\alpha/\ell_{s_n})K_0,$$

for all x, $a \le x \le b$, may be considered approximately to be the confidence region for $r(x)$. From (3.24) the region $\tilde{\tilde{G}}_\alpha(K)$ and the curves $\tilde{\tilde{h}}_n^{+}(x)$ and $\tilde{\tilde{h}}_n^{-}(x)$ are determined analogously. These curves are given by

$$\tilde{\tilde{h}}_n^{\pm}(x) = m_n(x) \pm \lambda'_\alpha \sqrt{q_n(x) - m_n^2(x) + \lambda'_\alpha q_n(x)} \,/(1 + \lambda'^2_\alpha),$$

where a non-negative quantity appears under the square root sign for large n.

We now state a theorem whose proof is analogous to the proof of Theorem 3.8.

THEOREM 3.9. Let $g(x)$, $\phi(x)$ and $V(x)$ satisfy the conditions of Theorem 3.7. If in addition to conditions on a_n and s_n one also requires the condition $na_n^{-2}/(\log s_n)^4 \to \infty$ then (3.23) and (3.24) in which $m_n(x)$ is replaced by the function $\tilde{m}_n(x)$ are valid.

Using this theorem one can determine in the same manner as it was done in the case of Theorem 3.8 a confidence region for $r(x)$.

The conditions of Theorem 3.8 regarding a_n and s_n are fulfilled if one, for example, sets $a_n = a_0 n^\alpha$ and $s_n = s_0 n^\beta$ for $1/3 < \alpha < 1/2$, $(1 - \alpha)/2 < \beta < \alpha$.

II. Testing Hypothesis Concerning a Regression Curve

b_1. Let $\sigma(x)$ be known and it is required to test the hypothesis H_0 that $r(x) = r_0(x)$ ($r_0(x)$ is a given function). The critical region for testing this hypothesis is determined approximately in accordance with (3.13) (or 3.20) by the inequality $M_n^* \geq d_n(\alpha)$ (or $\tilde{\tilde{M}}_n > d_n(\alpha)$) where

$$d_n(\alpha) = (na_n^{-1})^{1/2}(\ell_{s_n} + \lambda_\alpha/\ell_{s_n})K_0,$$

$$\lambda_\alpha = -\log|\log(1-\alpha)| + \log 2.$$

If, however, $\sigma(x)$ is unknown then the critical region is determined in view of (3.23) (or 3.24) by the inequality $\bar{M}_n \geq d_n(\alpha)$ (or $\bar{\bar{M}}_n \geq d_n(\alpha)$, where $\bar{\bar{M}}_n$ is the maximal deviation appearing in (3.24)).

b_2. Let it be required to test the hypothesis H_0 under the assumption that the regression curve is constant: $r(x) \equiv E(Y)$, $a \leq x \leq b$. If $\sigma(x)$ is known then the statistic

$$\ell_{s_n}\left(\frac{n}{a_n}\right)^{1/2} \max_{a\leq x\leq b}\left|m_n(x) - \overline{Y_n}\right|\sigma^{-1}(x) - \ell_{s_n}^2,$$

where

$$\overline{Y_n} = 1/n \sum_{i=1}^{n} Y_i$$

in view of Theorem 3.4 is asymptotically distributed in the same manner as M_n^*. Thus the hypothesis H_0 is rejected if

$$\max_{a\leq x\leq b}\left|m_n(x) - \overline{Y_n}\right|\sigma^{-1}(x) > d_n(\alpha).$$

Assume now that $\sigma(x)$ is unknown. In this case, in view of (3.23) the critical region for testing the hypothesis H_0 is determined in a similar manner with $\sigma(x)$ being replaced by its estimator $\sigma_n(x)$.

4. Limiting Distribution of Quadratic Deviation of Estimators of Regression Curves

1. In Section 2 of this chapter we introduced the quantity

$$U_n = E\int (r_n(x) - r(x))^2 g_n^2(x)h(x)\,dx$$

as a 'measure of the quality' of an estimator $r_n(x)$ and studied the rate of decrease of U_n to zero as $n \to \infty$.

In this section we shall investigate the limiting behavior of the distribution of the quadratic deviation

$$W_n = na_n^{-1} \int_I (r_n(x) - r(x))^2 g_n^2(x)\, h(x)\, dx,$$

where the integral is taken over a finite interval I. Without loss of generality we shall assume for simplicity that $I = [0, 1]$.

Our method of deriving the limiting distribution of the functional W_n is in essence based on an additional randomization [82] which is that instead of the n observations a random number of observations $\nu = \nu(n)$ is taken, distributed in accordance with the Poisson distribution with mean n and which is independent of the observations (X_i, Y_i), $i = 1, 2, \ldots$. This device allows us to reduce the study of the limiting distribution of the quadratic functional W_n to determination of the limiting distribution of a normalized sum of m-dependent random variables. This procedure was also used in paper [50]. The idea of the proofs of the results below is partially based on the arguments presented in [56].

Assumptions: 1°. $\min_{0 \le x \le 1} g(x) > 0$, $\min_{0 \le x \le 1} \beta(x) > 0$, $\beta(x) = D(Y|X = x)$, the density $g(x)$ is bounded on the whole axis. The random variable Y is bounded a.s.

2°. Functions $g(x)$ and $\phi(x)$ possess bounded derivatives up to the second order while $V(x)$ only possesses a bounded derivative of the first order.

3°. $K(x)$ is a symmetric kernel which equals zero outside the interval $[-1/2, 1/2]$,

$$\int K(x)\, dx = 1 \quad \text{and} \quad x^2 K(x) \in L_1(-\infty, \infty).$$

4°. $h(x)$ is a positive bounded and integrable function on I.

LEMMA 4.1. Let $na_n^{-4} \to 0$ as $n \to \infty$. Then

$$W_n - S_n = O_p(a_n^{-1/2}), \tag{4.1}$$

where

$$S_n = na_n^{-1} \int (\eta_n(x) - E\eta_n(x))^2 h(x)\, dx,$$

$$\eta_n(x) = \frac{a_n}{n} \sum_{i=1}^{n} \overline{\eta_i}(x), \qquad \overline{\eta_i}(x) = (Y_i - r(x)) K(a_n(x - X_i)).$$

Proof. We have

$$W_n - S_n = 2na_n^{-1} \int \eta_n(x) E\eta_n(x) h(x)\, dx - na_n^{-1} \int (E\eta_n(x))^2 h(x)\, dx = $$
$$= A_1 + A_2. \tag{4.2}$$

In view of conditions 2° and 3° we obtain that

$$|E\eta_n(x)| \le C_1 a_n^{-2}, \tag{4.3}$$

and hence

$$A_2 = na_n^{-1} \int (E\eta_n(x))^2 h(x)\ dx = O(na_n^{-5}). \tag{4.4}$$

Next since

$$\mathrm{cov}\ (\eta_n(x),\quad \eta_n(x_1)) = 1/na_n^2 (E(Y - r(x))K(a_n(x - X)) \times$$

$$\times (Y - r(x_1))K(a_n(x_1 - X)) - E\ \overline{\eta_1}(x) \cdot E\overline{\eta_1}(x_1)),$$

it follows that

$$E|A_1| \le 2na_n^{-1} \left(\frac{a_n^2}{n} \iint \left[\iint (y - r(x))(y - r(x_1))K(a_n(x - u))K \times \right.\right.$$

$$\left. \times (a_n(x_1 - y))E\eta_n(x_1)E\eta_n(x)h(x_1)h(x)\ dx\ dx_1 \right]^2 \times$$

$$\left. \times f(u,\ y)\ du\ dy \right)^{1/2} \le C_2 \sqrt{n}\ a_n^{-3}. \tag{4.5}$$

Thus we have from (4.2), (4.4) and (4.5) that

$$a_n^{1/2}\ E|W_n - S_n| \le C_3 \sqrt{n}\ a_n^{-5/2} + C_4 na_n^{-9/2} = O(1).$$

The lemma is thus proved.

Let $\nu(n)$ be a Poisson random variable with mean n and let $\nu(n)$ be independent of $(X_i,\ Y_i)$, $i = 1, 2, \ldots$.

Set

$$\eta_n^*(x) = n^{-1} a_n \sum_{j=1}^{\nu(n)} (Y_j - r(x))K(a_n(x - X_j)),$$

$$S_n^* = na_n^{-1} \int [\eta_n^*(x) - E\eta_n^*(x)]^2 h(x)\ dx.$$

Simple calculations show that

$$E\eta_n^*(x) = E\eta_n(x)$$

LEMMA 4.2. Let $na_n^{-2} \to \infty$ as $n \to \infty$. Then

$$E[\eta_n^*(x) - \eta_n(x)]^2 \le C_5 \frac{a_n}{n^{3/2}} + C_6 \frac{1}{na_n^4} \tag{4.6}$$

and

$$S_n - S_n^* = O_p(a_n^{-1/2}). \tag{4.7}$$

Proof. Centering $\eta_n(x)$ by the mathematical expectation $E\eta_n(x)$ and $\eta_n^*(x)$ by the conditional mathematical expectation $E(\eta_n^*(x)|\nu(n))$ we obtain

$$E[\eta_n^*(x) - \eta_n(x)]^2 = n^{-2}a_n^2 E\left[\sum_{i=\nu}^{n,} \alpha_i(x)\right]^2 + E(E\eta_n^*/\nu) -$$

$$- E\eta_n(x))^2 = B_1 + B_2 \tag{4.8}$$

where $\sum_{i=\nu}^{n}{}'$ denotes $\sum_{i=1}^{n} - \sum_{i=1}^{\nu}$ and $\alpha_i(x) = \overline{\eta_i}(x) - E\overline{\eta_i}(x)$.

It is easy to note that

$$B_2 = n^{-2}(a_n E\,\overline{\eta_1}(x))^2\, E(\nu - n)^2 = n^{-1}(a_n E\,\overline{\eta_1}(x))^2.$$

Whence in view of (4.3) we obtain

$$B_2 \leq C_7 n^{-1} a_n^{-4}. \tag{4.9}$$

We now bound B_1. We have

$$B_1 = E\left[n^{-1} a_n \sum_{j=m}^{n} \alpha_i(x)\right]^2 = n^{-2} a_n^2 |n - m| E\alpha_1^2(x) \leq$$

$$\leq C_8 n^{-2} a_n |n - m|, \tag{4.10}$$

since $g(x)$ is bounded and $E\alpha_1^2(x) \leq C_8 a_n^{-1}$. Now applying the unconditional expectation we arrive from (4.10)

$$B_1 \leq C_9 \frac{a_n}{n^{3/2}}. \tag{4.11}$$

Thus relations (4.8), (4.9) and (4.11) imply the validity of the bound (4.6).

We now proceed to prove (4.7). Note that

$$S_n - S_n^* = n a_n^{-1} \int [\eta_n(x) - \eta_n^*(x)]^2 h(x)\, dx + 2a_n^{-1} n \int (\eta_n^*(x) -$$

$$- E\eta_n^*(x))(\eta_n(x) - \eta_n^*(x)) h(x)\, dx. \tag{4.12}$$

Denote the first integral in the sum appearing on the r.h.s. of (4.12) by I_1 and the second by I_2.

Inequality (4.6) implies

$$EI_2 \leq C_{10} n a_n^{-1}\left(\frac{a_n}{n^{3/4}} + \frac{1}{n a_n^4}\right) = 0(a_n^{-1/2}). \tag{4.13}$$

It is easy to verify that

$$E[\eta_n^*(x) - E\eta_n^*(x)]^2 = \frac{a_n^2}{n} E[\,\overline{\eta_1}(x) - E\,\overline{\eta_1}(x)]^2$$

and by assumptions 1° and 3°

$$E[\,\overline{\eta_1}(x) - E\,\overline{\eta_1}(x)]^2 \leq C_{11} \int K^2(a_n(x - u) g(u)\, du \leq C_{12} a_n^{-1}.$$

Therefore

$$E[\eta_n^*(x) - E\eta_n^*(x)]^2 \le C_{13} \frac{a_n}{n} . \tag{4.14}$$

We thus deduce from here and from (4.6) that

$$EI_2 \le C_{14} a_n^{-1} n \left(\frac{a_n}{n}\right)^{1/2} \left(\frac{a_n}{n^{3/2}} + \frac{1}{na_n^4}\right)^{1/2} =$$

$$= C_{14} n^{-1/4} + a_n^{-5/2} = (a_n^{-1/2}). \tag{4.15}$$

Whence (4.12)-(4.15) imply (4.7). The lemma is thus proved.

Denote by $F_\nu(x, y)$ the empirical distribution function of the sample $(X_1, Y_1), \ldots, (X_\nu, Y_\nu)$ and let $\hat{F}_n(x, y) = \frac{\nu}{n} F_\nu(x, y)$. Then $n\hat{F}_n(x, y)$ will be a Poisson process on the plane and

$$S_n^* = \int [\sqrt{na_n} \int\int (y - r(x))K(a_n(x - u)) \, d(\hat{F}_n(u, y) -$$

$$- F(u, y))]^2 h(x) \, dx,$$

where $F(x, y)$ is the distribution function of (X, Y).

We subdivide the interval $[0, 1]$ into $N = [a_n]$ parts Δ_j, $j = \overline{1, N}$, of length a_n^{-1} possibly with a small remainder ($[a]$ denotes the integer part of a) and consider the sum

$$\hat{S}_n = \sum_{j=1}^{N} \xi_j,$$

where

$$\xi_j = \int_{\Delta_j} [\sqrt{na_n} \int\int (y - r(x))K(a_n(x - u)) \, d(\hat{F}_n(u, y) -$$

$$- F(u, y))]^2 h(x) \, dx,$$

$$\Delta_j = [(j - 1)a_n^{-1}, \quad ja_n^{-1}), \quad j = \overline{1, N}.$$

Note that $\xi_1, \xi_2, \ldots, \xi_N$ are $m = 1$-dependent random variables, i.e. if $|k - j| > 1$ then ξ_k and ξ_j are independent. This follows immediately from the fact that $n\hat{F}_n$ is a Poisson process and the kernel $K(x) = 0$ outside the interval $[- 1/2, 1/2]$.

From (4.14) we have

$$E|S_n^* - \hat{S}_n| \le C_{15}(1 - [a_n]a_n^{-1}) \le C_{15} a_n^{-1}.$$

Hence

$$S_n^* - \hat{S}_n = O_p(a_n^{-1/2}). \tag{4.16}$$

Set

$$V_j = \xi_j - E\xi_j, \quad \sigma_n^2 = ND\left(\sum_{j=1}^{N} V_j\right).$$

Then

$$\hat{S} - E\hat{S}_n = \sum_{j=1}^{N} V_j \quad \text{and} \quad \sigma_n^2 = N(E\,\hat{S}_n^2 - (E\,\hat{S}_n)^2,$$

and from the property of a stochastic integral over a Poisson process and assumption 2° we have

$$E\,\hat{S}_n = a_n \int \left(\int\int (y - r(x))^2 K^2(a_n(x - u)) f(u, y)\,du\,dy \right) \times$$

$$\times\, h(x)\,dx = \int \alpha(x) h(x)\,dx \int K^2(x)\,dx + O(a_n^{-1}), \tag{4.17}$$

where

$$\alpha(x) = D(Y|X = x) g(x).$$

We shall now study the asymptotics of $E\,\hat{S}_n^2$. After a simple transformation we obtain

$$E\,\hat{S}_n^2 = E \int\int \left\{ \left[\sqrt{\frac{a_n}{n}} \sum_{i=1}^{\nu} \alpha_i(x_1) + \sqrt{na_n}\left(\frac{\nu}{n} - 1\right) E\,\overline{\eta_1}(x_1) \right] \times \right.$$

$$\left. \times \left[\sqrt{\frac{a_n}{n}} \sum_{j=1}^{\nu} \alpha_j(x_2) + \sqrt{na_n}\left(\frac{\nu}{n} - 1\right) E\,\overline{\eta_1}(x_2) \right] \right\}^2 \times$$

$$\times\, h(x_1) h(x_2)\,dx_1\,dx_2\,. \tag{4.18}$$

We introduce the notation

$$a(x_i) = \sqrt{\frac{a_n}{n}} \sum_{j=1}^{\nu} \alpha_j(x_i), \quad i = 1, 2,$$

$$b(x_i) = \sqrt{na_n}\left(\frac{\nu}{n} - 1\right) E\,\overline{\eta_1}(x_i), \quad i = 1, 2.$$

$$I(k_1, k_2, k_3, k_4) = \int E a^{k_1}(x_1) a^{k_2}(x_2) b^{k_3}(x_1) b^{k_4}(x_2) h(x_1)$$

$$h(x_2)\,dx_1\,dx_2.$$

Then it follows from (4.18) that

$$E\,S_n^2 = I(2, 2, 0, 0) + I(2, 0, 0, 2) + I(0, 2, 2, 0) +$$

$$+ I(0, 0, 2, 2) + 2I(2, 1, 1, 0) + 2I(0, 1, 2, 1) +$$

$$+ 2I(1, 2, 1, 0) + 2I(1, 0, 1, 2) + 4I(1, 1, 1, 1). \tag{4.19}$$

Consider the integral I(2, 2, 0, 0). Simple calculations show that

$$I(2, 2, 0, 0) = a_n^2 (\frac{1}{n} - 1) \int E\alpha_1^2(x_1)\alpha_2^2(x_2)h(x_1)h(x_2)\, dx_1\, dx_2 +$$

$$+ 2a_n^2 \int (E\alpha_1(x_1)\alpha_1(x_2))^2 h(x_1)h(x_2)\, dx_1\, dx_2 =$$

$$= I_1 + I_2.$$

It is easy to verify that $I_1 = O(a_n^{-1}) + \Delta^2$ and

$$I_2 = 2a_n^2 \int [\int K(a_n(x_1 - u))K(a_n(x_2 - u))\alpha(u)\, du]^2 h(x_1)h(x_2) \times$$

$$\times dx_1\, dx_2 + O(a_n^{-2}) \cong 2a_n^{-1} \int \alpha^2(x)h^2(x)\, dx \int K_0^2(u)\, du,$$

where

$$K_0(x) = \int K(t)K(x - t)\, dt, \quad \Delta = \int \alpha(x)h(x) \int K^2(u)\, du.$$

Thus

$$I(2, 2, 0, 0) \cong 2a_n^{-1} \int \alpha^2(x)h^2(x)\, dx \int K_0^2(u)\, du + \Delta^2. \tag{4.20}$$

As far as the remaining integrals in (4.19) are concerned, it is easy to establish that they are all $O(a_n^{-1})$. To verify this, estimates of the moments of the random variable ν-n are required. Utilizing the decomposition for the characteristic function of the variable ν-n

$$\exp\, (it(\nu - n)) = \exp \left(\frac{(it)^2}{2!} n + \frac{(it)^3}{3!} n + \ldots \right)$$

and the relation between the moments and the cumulants of the random variable ν-n [58] we arrive at the following bounds on the moments of the random variable ν-n:

$$E|\nu - n|^\ell \leq C(\ell)n^{\ell/2}, \quad \ell \geq 2 \tag{4.21}$$

Next, consider for example I(0, 0, 2, 2) and I(1, 1, 1, 1). (The remaining integrals are estimated analogously).

Utilizing (4.21) and $E\, \overline{\eta_1}(x) = O(a_n^{-3})$ we obtain

$$I(0, 0, 2, 2) = n^2 a_n^2 \iint E\, (\frac{\nu}{n} - 1)^4 (E\, \overline{\eta_1}(x_1))^2 (E\, \overline{\eta_1}(x_2))^2 \times$$

$$\times\, h(x_1)h(x_2)\, dx_1\, dx_2 = O(a_n^{-10}),$$

and

$$I(1, 1, 1, 1) = a_n^2 \iint E\alpha_1(x_1)\alpha_1(x_2)E\nu(\frac{\nu}{n} - 1)^2 E\, \overline{\eta_1}(x_1) \times$$

$$\times\, E\, \overline{\eta_1}(x_2)h(x_1)h(x_2)\, dx_1\, dx_2 \leq C_{16} n a_n^{-5} = O(a_n^{-1}).$$

Thus (4.17) and (4.20) imply that

$$\lim_{n\to\infty} N[E\hat{S}_n^2 - (E\hat{S}_n)^2] = 2 \int \alpha^2(x)h^2(x)\,dx \int K_0^2(u)\,du.$$

Whence

$$\lim_{n\to\infty} \sigma_n^2 = 2 \int \alpha^2(x)h^2(x)\,dx \int K_0^2(u)\,du \tag{4.22}$$

We shall now bound $E|V_j|^\ell$, $\ell \geq 2$. We have

$$E|V_\ell|^\ell \leq 2^{\ell-1}(E|\xi_j|^\ell + |E\xi_j|^\ell). \tag{4.23}$$

However

$$E\xi_i = \int_{\Delta_i} \alpha(x)h(x)\,dx + O(a_n^{-2}) = O(a_n^{-1})$$

Hence

$$|E\xi_j|^\ell = O(a_n^{-\ell}). \tag{4.24}$$

Analogously to (4.18) we have

$$\vec{E}|\xi_j|^\ell \leq 2^{2\ell-1}\left\{E\left(\int_{\Delta_j}\left(\sqrt{\frac{a_n}{n}}\sum_{i=1}^{\nu}\alpha_i(x)\right)^2 h(x)\,dx\right)^\ell + E\left(\int_{\Delta_j} na_n \times \right.\right.$$
$$\left.\left. \times \left(\frac{\nu}{n} - 1\right)^2 (E\,\overline{\eta_1}(x)^2 h(x)\,dx\right)^\ell\right\} = A_1 + A_2. \tag{4.25}$$

Whence in view of (4.21)

$$A_2 \leq C_{17}(\ell)a_n^{-\ell} \tag{4.26}$$

and also in view of Hölder's inequality

$$A_1 \leq C_{18}(\ell) \int_{\Delta_j} E\left(\sqrt{\frac{a_n}{n}}\sum_{i=1}^{\nu}\alpha_i(x)\right)^{2\ell} dx\; a_n^{-(\ell-1)}. \tag{4.27}$$

Furthermore we have

$$a_n^\ell n^{-\ell} E\left(\sum_{i=1}^{m}\alpha_i(x)\right)^{2\ell} = \left(\frac{m}{n} D\beta_1(x)\right)^\ell E\left(m^{-1/2}\sum_{i=1}^{m}\frac{\beta_i(x)}{\sqrt{D\beta_1(x)}}\right)^{2\ell}. \tag{4.28}$$

where

$$\beta_i(x) = \sqrt{a_n}\,\alpha_i(x),$$

and moreover

$$D\beta_i(x) \cong \alpha(x) \int K^2(u)\,du.$$

From von Bahr's formula (1.1.21) we have

$$E\left(m^{-1/2}\sum_{i=1}^{m}\frac{\beta_i(x)}{\sqrt{D\beta_1(x)}}\right)^{2\ell} = \int x^{2\ell}\,d\Phi(x) + \sum_{j=1}^{2\ell-2} m^{-j/2}\int u^{2\ell}\,\times$$

$$\times\, dP_j(-\Phi)(u), \tag{4.29}$$

where $P_j(-\Phi)$ are determined by formula (1.1.22). It is easy to verify that the cumulants γ_i, $i \geq 3$ appearing in $P_j(-\Phi)$ satisfy the inequality

$$|\gamma_i| \leq C_{19}(a_n/D\beta_1(x))^{(i-2)/2}, \quad i \geq 3$$

so that

$$\left|\int u^{2\ell}\,dP_j(-\Phi)(u)\right| \leq C_{20}(a_n/D\beta_1(x))^{j/2}.$$

From here taking (4.29) into account we have

$$E(m^{-1/2}\sum_{j=1}^{m}\beta_j(x)/\sqrt{D\beta_1(x)}\,)^{2\ell} \leq \int x^{2\ell}\,d\Phi(x) + C_{21}\sum_{k=1}^{2\ell-2} m^{-k/2}\,\times$$

$$\times\,(a_n/D\beta_1(x))^{k/2}.$$

In view of this relationship one concludes from (4.28) that

$$\int_{\Delta_j} E\left(\sqrt{\frac{a_n}{n}}\sum_{j=1}^{m}\alpha_j(x)\right)^{2\ell} dx \leq n^{-\ell}\left[m^{\ell}\int_{\Delta_j}(D\beta_1(x))^{\ell}\,dx + C_{21}\,\times\right.$$

$$\left.\times\sum_{j=1}^{2\ell-2} m^{\ell-(j/2)}a_n^{j/2}\int_{\Delta_j}(D\beta_1(x))^{\ell-(j/2)}\,dx\right] \leq C_{22}n^{-\ell}\,\times$$

$$\times\, a_n^{-1}\left[m^{\ell} + C_{21}\sum_{j=1}^{2\ell-2} m^{\ell-(j/2)}a_n^{j/2}\right]. \tag{4.30}$$

Taking the unconditional mathematical expectation and utilizing the inequality (4.21) we obtain from (4.30) that

$$\int_{\Delta_j} E\left(\sqrt{\frac{a_n}{n}}\sum_{i=1}^{\nu}\alpha_i(x)\right)^{2\ell} dx \leq C_{23}a_n^{-1}\left(C_{24} + C_{25}\sum_{j=1}^{2\ell-2}\,\times\right.$$

$$\left.\times\left(\frac{a_n}{n}\right)^{j/2}\right) \leq C_{26}a_n^{-1}.$$

Whence in view of (4.26)

$$A_1 \leq C_{27}a_n^{-\ell}\,. \tag{4.31}$$

Thus substituting inequalities (4.26) and (4.31) into (4.25) and then (4.24) and (4.25) into (4.23) we obtain that

$$E|V_j|^{\ell} \le C_{28} a_n^{-\ell}, \quad \ell \ge 2. \tag{4.32}$$

We now proceed to an investigation of the limiting distribution of the random variable $\sqrt{a_n}\,(\hat{S} - E\hat{S}_n)$. Here we are going to use the method of the proof of the central limit theorem of Hoeffding and Robbins for the case of m-dependent random variables [59].

Set

$$U_i = V_{(i-1)k+1} + \ldots + V_{ik-1}, \quad i = 1, 2, \ldots, n_1$$

$$T_n = \sum_{j=1}^{n_1-1} V_{jk} + V_{n_1 k} + \ldots + V_N,$$

where $k = [N^{\alpha}]$, $0 < a < 1/4$ and $n_1 = [\frac{N}{K}]$ so that

$$N = n_1 k + r, \quad 0 < r < k.$$

In view of inequality (4.32) we have

$$a_n ET_n^2 \le (n_1 - 1)a_n^{-1} + (k + 1)^2 a_n^{-1} = O(1), \tag{4.33}$$

$$\sum_{i=1}^{n_1} E|k^{1/2}U_i|^3 = k^{3/2} \sum_{i=1}^{n_1} E\left|\sum_{j=1}^{k-1} V_{(i-1)k+j}\right|^3 \le C_{29} a_n^{-(4-7\alpha)/2} \tag{4.34}$$

and some simple manipulations yield

$$\left|n_1 \sum_{i=1}^{n_1} E(k^{1/2}U_i)^2 - \sigma_n^2\right| \le C_{30} n_1 (n_1 k - a_n)(k - 1)a_n^{-2} +$$
$$+ C_{31} n_1 a_n^{-1} + (N - n_1 k)a_n^{-1} = O(1).$$

From here and from (4.22) we obtain that

$$\lim_{n\to\infty} n_1 \sum_{i=1}^{n_1} E(k^{1/2}U_i)^2 = 2\int \alpha^2(x)h^2(x)\,dx \int K_0^2(u)\,du. \tag{4.35}$$

The normalized sum

$$\sum_{i=1}^{n_1} k^{1/2}U_i) / \left(\sum_{i=1}^{n_1} E(k^{1/2}U_i)^2\right)^{1/2}$$

is distributed asymptotically normally (0, 1) since for a sequence of independent random variables $\{k^{1/2}U_i,\ i = 1, 2, \ldots\}$ in view of (4.34) and (4.35) the Lyapunov condition is fulfilled, namely

$$\sum_{i=1}^{n_1} E|k^{1/2}U_i|^3 / \left[\sum_{i=1}^{n_1} E(k^{1/2}U_i)^2\right]^{3/2} \le C_{32} a_n^{-((1-4\alpha)/2)}. \tag{4.36}$$

Note that

$$\sqrt{a_n}\left(\hat{S}_n - E\hat{S}_n\right) = \sqrt{a_n}\, T_n + \frac{\sqrt{a_n}}{n_1^{1/2} k^{1/2}}\, n_1^{1/2} \sum_{i=1}^{n_1} k^{1/2} U_i,$$

and moreover $\sqrt{a_n}\, n_1^{-1/2} k^{-1/2} \to 1$ as $n \to \infty$. In view of (4.36) $n_1^{1/2} \sum_{i=1}^{n_1} \times$ $\times\, k^{1/2} U_i$ is asymptotically distributed normally $(0, \sigma)$ where

$$\sigma^2 = \int \alpha^2(x) h^2(x)\, dx \int K_0^2(u)\, du,$$

and in view of (4.33) $\sqrt{a_n}\, T_n \to 0$ in probability. Thus, $\sqrt{a_n}\, (\hat{S}_n - E\hat{S}_n)$ is distributed in the limit normally with the same parameters $(0, \sigma)$.

Thus taking (4.7), (4.16) and (4.17) into account we arrive at the basic

THEOREM 4.1. Let the assumptions 1°-4° be fulfilled. If $na_n^{-2} \to \infty$ and $na_n^{-4} \to 0$ then the random variable

$$\sqrt{a_n}\, (W_n - \int \alpha(x) h(x)\, dx \int K^2(u)\, du)$$

is asymptotically normally distributed with mean 0 and variance

$$\sigma^2 = 2 \int \alpha^2(x) h^2(x)\, dx \int K_0^2(u)\, du, \qquad K_0 = K{*}K.$$

Remark. An analogous theorem was later proved in [60] using another method which was developed in [61]. However the conditions imposed on a_n in this theorem are quite rigid.

2. We shall now consider statistical problems associated with regression curves.

I. Utilizing the result of Theorem 4.1 we can construct tests for testing hypotheses concerning the regression curve. Let $\alpha(x)$ be known and it is required to test the null hypothesis H_0 that $r(x) = r_0(x)$ (where $r_0(x)$ is a given function). To accomplish this it is necessary to calculate W_n for $r(x) = r_0(x)$ and reject the hypothesis if $W_n \geq d_n(\alpha)$ where

$$d_n(\alpha) = \int \alpha(x) h(x)\, dx \int K^2(x)\, dx + a_n^{-1/2} \lambda_\alpha \sigma,$$

λ_α is the quantile of level α of the standard normal distribution.

II. Let $\alpha(x)$ be unknown. Then for testing the hypothesis H_0 one must have an analog of Theorem 4.1 where $\alpha(x)$ appearing in the normalizing constant is replaced by

$$\alpha_n(x) = V_n(x) - r_0^2(x) g_n(x) = a_n/n \sum_{j=1}^{n} (Y_j^2 - r_0^2(x)) K(a_n(x - X_j)).$$

Set

$$\Delta_n = \int \alpha_n(x) h(x) \int K^2(u)\, du, \quad \Delta = \int \alpha(x) h(x)\, dx \int K^2(u)\, du,$$

$$\sigma_n^2 = \int \alpha_n^2(x) h^2(x)\, dx \int K_0^2(x)\, dx.$$

We have

$$E|\Delta_n - \Delta| \leq \int E|V_n(x) - V(x)|\, h(x)\, dx + C_{33} \int E|g_n(x) - g(x)|\, h(x)\, dx,$$

moreover, in view of assumptions 1° and 2°

$$E|V_n(x) - V(x)|^2 \leq C_{34} a_n/n + C_{35} a_n^{-2},$$

$$E|g_n(x) - g(x)|^2 \leq C_{36} a_n/n + C_{37} a_n^{-2}.$$

Whence

$$\sqrt{a_n}\, E|\Delta_n - \Delta| \leq C_{38} \left(\frac{a_n^2}{n} + a_n^{-1} \right)^{1/2} = O(1). \tag{4.37}$$

Analogously

$$E|\sigma_n^2 - \sigma^2| \leq C_{39} \left(\frac{a_n}{n} \right)^{1/2} = O(1). \tag{4.38}$$

Taking (4.37) and (4.38) we arrive from Theorem 4.1 at

THEOREM 4.2. Under the conditions of Theorem 4.1 the random variable

$$\sqrt{a_n}\, \sigma_n^{-1} (W_n - \Delta_n)$$

is asymptotically normally distributed with mean 0 and variance 1.

This theorem allows us to construct a level α test for testing the hypothesis $H_0: r(x) = r_0(x)$. Here the hypothesis H_0 is rejected if $W_n \geq \tilde{d}_n(\alpha) = \Delta_n + a_n^{-1/2} \lambda_\alpha \sigma_n$.

Consider now a sequence of 'close' alternatives to the hypothesis H_0 of the form

$$H_n: \overline{r_n}(x) = r_0(x) + \gamma_n \bar{r}(x) + O(\gamma_n), \quad \gamma_n = n^{-(1/2)+\delta/4} \tag{4.39}$$

where $O(\gamma_n)$ is uniformly in x. (Alternatives which differ from the null hypothesis by a quantity of order $\gamma_n = n^{-(1/2)+\delta/4}$ were also considered in Section 2 of Chapter 3.) Clearly, the regression curves of the form (4.39) correspond to the class of densities of the form

$$\bar{f}_n(x, y) = f(x, y) + \gamma_n \bar{f}(x, y) + O(\gamma_n).$$

Let $\bar{g}_n(x)$, $\bar{\phi}_n(x)$ and $\bar{V}_n(x)$ denote $g(x)$, $\phi(x)$ and $V(x)$ respectively, evaluated relative to the density $\bar{f}_n(x, y)$.

THEOREM 4.3. Let $\bar{g}_n(x)$, $\bar{\phi}_n(x)$ and $\bar{V}_n(x)$ satisfy the conditions of

Theorem 4.1. If $a_n = n^\delta$, $1/4 < \delta < 1/2$ then as $n \to \infty$ we have

(a) $$P_{H_n}(W_n \geq d_n(\alpha)) \to 1 - \phi\left(\lambda_\alpha - \frac{1}{\sigma}\int \bar{r}^2(x)g^2(x)h(x)\,dx\right);$$

(b) $$P_{H_n}(W_n \geq \tilde{d}_n(\alpha)) \to 1 - \phi\left(\lambda_\alpha - \frac{1}{\sigma}\int \bar{r}^2(x)g^2(x)h(x)\,dx\right).$$

Proof: We have

$$W_n = \frac{n}{a_n}\int (\phi_n(x) - \bar{r}_n(x)g_n(x))^2 + 2\frac{n}{a_n}\int (\phi_n(x) - \bar{r}_n(x) \times$$

$$\times g_n(x))(\bar{r}_n(x) - r_0(x))g_n(x)h(x)\,dx + \frac{n}{a_n}\int (\bar{r}_n(x) -$$

$$- r_0(x))^2 g_n^2(x)h(x)\,dx = W_n^{(1)} + W_n^{(2)} + W_n^{(3)}. \quad (4.40)$$

In view of Theorem 4.1 the random variable $a_n^{1/2}(W_n^{(1)} - \Delta)$ is distributed in the limit normally $(0, \sigma)$. Furthermore, it follows from (4.39) for our choice of a_n and γ_n that

$$a_n^{1/2}W_n^{(3)} = \int \bar{r}^2(x)g^2(x)h(x)\,dx + o_p(1).$$

We bound $a_n^{1/2}W_n^{(2)}$. Since $E(a_n(Y - \bar{r}_n(x))K(a_n(x - X)) = O(a_n^{-2})$ we have

$$a_n^{1/2}W_n^{(2)} = n^{1/2}\,a_n^{-1/2}\,n^{\delta/4}\int\left(\frac{a_n}{n}\sum_{j=1}^{n}\alpha_j(x)\right)g_n(x)\bar{r}(x)h(x)\,dx +$$

$$+ O_p(n^{(2-9\delta)/4}),$$

where $\alpha_j(x) = \eta_j(x) - E\eta_j(x)$, $\eta_j(x) = (Y_j - \bar{r}_n(x))K(a_n(x - X_j))$.

In view of (4.1.3) and (4.1.6)

$$E\left(\frac{a_n}{n}\sum_{j=1}^{n}\alpha_j(x)\right)^2 \leq C_{40}a_n/n$$

and

$$E(g_n(x) - Eg_n(x))^2 \leq C_{41}a_n/n.$$

Therefore we obtain from (4.41) that

$$a_n^{1/2}W_n^{(2)} = (n/a_n)^{1/2}n^{\delta/4}\int\left[\frac{a_n}{n}\sum_{j=1}^{n}\alpha_j(x)\right]Eg_n(x)\bar{r}(x)h(x)\,dx +$$

$$+ O_p(n^{(2-9\delta)/4}) + O_p(n^{(3\delta-2)/4}). \quad (4.42)$$

Next a simple calculation of the variance of the stochastic integral in (4.42) shows that $a_n^{1/2} W_n^{(2)} = O_p(1)$. Thus assertion (a) of the theorem is proved. Assertion (b) follows from (a) in an obvious manner.

3. Testing Hypothesis about Equality of two Regression Functions

Let two independent samples $(X_j^{(\ell)}, Y_j^{(\ell)})$, $j = 1, \ldots, n_i$, $i = 1, 2$ be given from the population $(X^{(i)}, Y^{(i)})$. Next, let $r_i(x)$, $i = 1, 2$, be the corresponding regression curves. Assume that marginal densities of X's are the same, i.e.

$$g_{X^{(1)}}(x) = g_{X^{(2)}}(x) = g(x), \quad \text{and} \quad \beta(x) = D(Y^{(1)}/X^{(1)} = x) =$$

$$= D(Y^{(2)}/X^{(2)} = x)$$

We shall choose $\hat{r}_i(x) = g_{n_1 n_2}^{-1}(x)\phi_{n_i}(x)$, $i = 1, 2$, as an estimator of $r_i(x)$.

Here

$$\phi_{n_i}(x) = \sum_{j=1}^{n_i} Y_j^{(i)} K(a_{n_i}(x - X_j^{(i)})) \frac{a_{n_i}}{n_i}, \quad g_{n_i}(x)$$

is a kernel-type density estimator and

$$g_{n_1 n_2}(x) = (n_1 + n_2)^{-1}(n_1 g_{n_1}(x) + n_2 g_{n_2}(x)).$$

Quite often it is required to test the hypothesis that regression functions are the same, i.e. to test the hypothesis $H_0: r_1(x) = r_2(x)$.

For comparing the estimators $\hat{r}_1(x)$ and $\hat{r}_2(x)$ we introduce the statistic

$$L_{n_1 n_2} = \frac{N_1 N_2}{N_1 + N_2} \int (\phi_{n_1}(x) - \phi_{n_2}(x))^2 h(x)\, dx, \quad N_i = n_i a_{n_i}^{-1}.$$

Let $n_1/n_2 = \tau$, $a_{n_1} = a_{n_2} = a_n$, $n = n_2(1 + \tau)$. Using the method of proof of Theorem 4.1 one can show that under H_0 the statistic $a_n^{1/2}(L_{n_1 n_2} - \Delta)$ is distributed asymptotically normally $(0, \sigma)$:

$$\Delta = \int V(x)h(x)\, dx \int K^2(u)\, du, \quad \sigma^2 = 2\int V^2(x)h^2(x)\, dx \int K_0^2(u)\, du,$$

$$V(x) = g(x)(D(Y^{(1)}/X^{(1)} = x) + r_1^2(x)).$$

This assertion allows us to construct an asymptotic test of level α for testing the hypothesis H_0:

$$r_1(x) \equiv r_2(x).$$

In this test the hypothesis is rejected if $L_{n_1n_2} \geq \Delta_{n_1n_2} + a_n^{-1/2}\sigma_{n_1n_2}\lambda_\alpha$. Here $\Delta_{n_1n_2}$ and $\sigma^2_{n_1n_2}$ are obtained from Δ and σ^2 by replacing $V(x)$ with

$$V_{n_1n_2}(x) = (n_1 + n_2)^{-1}(n_1V_{n_1}(x) + n_2V_{n_2}(x)),$$

where

$$V_{n_i}(x) = \frac{a_{n_i}}{n_i} \sum_{j=1}^{n_i} (Y_j^{(i)})^2 K(a_{n_i}(x - X_j^{(i)})), \quad i = 1, 2.$$

5. Nonparametric Estimates of Components of a Convolution (S. N. Bernstein's Problem)

In this section a situation is considered for which values of a certain random variable X are observed. Moreover, the variable X is assumed to be a sum of two independent variables Y and Z; Z has a normal distribution $N(0, \sigma^2)$ where σ is assumed to be known. The variable of interest is Y since Z can be viewed as a measurement error of Y. Given observations of variable X it is required: 1° to estimate the probability density of variable Y (S. N. Bernstein's problem [25]). 2° to estimate the regression function of Y on X, i.e. to estimate $r(x) = E(Y|X = x)$.

A solution to these problems - as we shall see below - is closely connected with an investigation of properties of estimators of derivatives of a probability density. It is therefore desirable to present first certain basic properties of these estimators. Although estimators of derivatives of densities are obtained naturally from the kernel-type estimators, there were few attempts in the literature to apply these estimators to statistical problems. A solution of the problems stated above may serve as an example of these applications.

1. Let X_i, $i = \overline{1, n}$, be a sample from n independent observations on a random variable X with the density function $f(x)$. Assume that $f(x)$ possesses bounded derivatives up to the s + 1-th order. Assume also that the kernel $K(x)$ possesses bounded derivatives up to the s + 1-th order and let $\lim_{|x|\to\infty} K^{(r)}(x) = 0$, $r = 0, 1, \ldots, s$, be symmetric, $xK(x)$ $L_1(-\infty)$, and

$$\int K(x)\, dx = 1.$$

It is natural to choose as an estimator of $f^{(s)}(x)$ the statistic

$$f_n^{(s)}(x) = n^{-1}a_n^{s+1} \sum_{j=1}^{n} K^{(s)}(a_n(x - X_j)). \tag{5.1}$$

THEOREM 5.1. $f_n^{(s)}(x)$ is an asymptotically unbiased estimator of $f_n^{(s)}(x)$ uniformly in x and moreover

$$\sup_{x \in R_1} |Ef_n^{(s)}(x) - f^{(s)}(x)| = O(a_n^{-1}). \tag{5.2}$$

Proof. The identity

$$a_n^{s+1} K^{(s)}(a_n(x - u))f(u) = a_n K(a_n(x - u))f^{(s)}(u) - \frac{d}{du} \sum_{j=0}^{s-1} \times$$

$$\times a_n^{j+1} K^{(j)}(a_n(x - u))f^{(s-j-1)}(u),$$

together with the fact that $K^{(s)}(x) \to 0$ as $|x| \to \infty$, $s = 0, 1, \ldots,$ yields that

$$Ef_n^{(s)}(x) = a_n^{s+1} \int K^{(s)}(a_n(x - u))f(u)\, du =$$

$$= a_n \int K(a_n(x - u))f^{(s)}(u)\, du. \tag{5.3}$$

whence

$$|Ef_n^{(s)}(x) - f^{(s)}(x)| \leq \max_{-\infty<x<\infty} |f^{(s+1)}(x)|\, a_n^{-1} \int |u||K(u)|\,du,$$

q.e.d.

THEOREM 5.2. Let $K^{(s)}(x)$ be a function of a bounded variation and let the series $\sum_1^n \exp\{- \gamma n a_n^{-2(s+1)}\} < \infty$ for any $\gamma > 0$, then

$$\sup_{-\infty<x<\infty} |f_n^{(s)}(x) - f^{(s)}(x)| \to 0 \quad \text{a.s. as} \quad n \to \infty.$$

The proof of this theorem essentially coincides with the proof of Theorem 2.1.1 and is therefore omitted.

The asymptotic normality of the estimator (5.1) is presented in Theorem 1.2.2.

2. Below the theorem due to Rooney [26] will be required. For reference convenience we now present the statement of this theorem.

THEOREM 5.3 [26]. Let $f_1(x)$ be the Gaussian transform of the function $f_2(x)$:

$$f_1(x) = \frac{1}{\sigma}\phi\left(\frac{x}{\sigma}\right) * f_2(x), \tag{5.4}$$

where

$$\phi(x) = \frac{1}{\sqrt{2\pi}} \exp\left\{-\frac{x^2}{2}\right\}.$$

If $f_2(x)$ is continuous and $\exp\{- ((x - y)^2/4\sigma^2)\}f_2(y) \in L_1(-\infty, \infty)$ then

$$\lim_{\tau \to 1-} \sum_{k=0}^{\infty} (-1)^k f_1^{(2k)}(x) \frac{(\sigma^2\tau)^k}{2^k k!} = f_2(x).$$

This theorem implies

LEMMA 5.1. Let $\{\tau_n\}$, $0 < \tau_n < 1$, be a sequence of numbers such that $\lim_{n\to\infty} \tau_n = 1$ and $\lim_{n\to\infty} \tau_n^{\nu(n)}/(1 - \tau_n) = 0$ where $\nu(n) \to \infty$ as $n \to \infty$ and $f_2(x)$ satisfies the condition of the theorem. Then

$$g_n(x) = \sum_{i=1}^{\nu(n)} A_i(n) f_1^{(2i)}(x) \to f_2(x) \quad \text{as} \quad n \to \infty \tag{5.5}$$

where

$$A_i(n) = (-1)^i (\sigma^2 \tau_n)^i / 2^i i!$$

Proof. We have

$$g_n(x) = \sum_{i=1}^{\infty} A_i(n) f_1^{(2i)}(x) - \sum_{i=\nu+1}^{\infty} A_i(n) f_1^{(2i)}(x). \tag{5.6}$$

In view of the theorem the first summand approaches $f_2(x)$. With the aid of the inequality

$$|H_m(x)| \leq C_1 2^{m/2} \sqrt{m!}\, e^{x^2/2}$$

where $H_m(x)$ is the Hermite polynomial, we obtain from (5.4) the inequality

$$|f_1^{(2k)}(x)| \leq C_1 \sqrt{2k!}\, /\sigma^{2k}. \tag{5.7}$$

This implies that the second sum in (5.6) converges to 0. The lemma is thus proved.

3. Let $X = Y + Z$ be a sum of independent random variables Y and Z and let the density Z be normal with mean 0 and known variance σ^2 and let Y possess an unknown density. Based on observed values $X_1, X_2, \ldots, X_n$ of the variable X it is required to estimate $g(x)$.

As an estimator of $g(x)$ we shall choose the statistic

$$\hat{g}_n(x) = \sum_{i=1}^{\nu(n)} A_i(n) f_n^{(2i)}(x),$$

where $f_n^{(2i)}(x)$ is an estimator of the type (5.1). Here it is assumed that

$$K(x) = \frac{1}{\sigma}\phi\left(\frac{x}{\sigma}\right) * K_1(x), \quad \text{where} \quad K_1(x) = \frac{1}{\sigma}\phi\left(\frac{x}{\sigma}\right) * K_0(x),$$

and $K_0(x)$ is a known probability density.

THEOREM 5.4. If $\tau_n^{\nu(n)}/(1 - \tau_n) \to 0$ and $a_n^{-1}(1 - \tau_n)^{-1} \to 0$ as $n \to \infty$ then

$\hat{g}_n(x)$ is an asymptotically unbiased estimator of $g(x)$.

Proof. In view of equation (5.3) we have

$$E\hat{g}_n(x) = \sum_{s=1}^{\nu(n)} A_s(n) f^{(2s)}(x) + a_n^{-1} \sum_{s=1}^{\nu(n)} A_s(n) \int uK(u) \times$$
$$\times f^{(2s+1)}(\xi)\, du,$$

where $f(x)$ is the density of X. Whence applying Lemma 5.1 and inequality (5.7) we obtain

$$E\hat{g}_n(x) = \sum_{s=1}^{\nu(n)} A_s(n) f^{(2s)}(x) + O(a_n^{-1}(1 - \tau_n)^{-1}) = g(x) + O(1). \tag{5.8}$$

q.e.d.

THEOREM 5.5. If

$$\tau_n^{\nu(n)}(1 - \tau_n)^{-1} \to 0, \quad a_n^2 n^{-1}(1 - \tau_n)^{-2} \to 0 \quad \text{and}$$

$a_n^{-1}(1 - \tau_n)^{-1} \to 0$ as $n \to \infty$, then $\hat{g}_n(x)$ is a consistent estimator of the density $g(x)$.

Proof. It is easy to verify that

$$E\hat{g}_n^2(x) = n^{-1} \sum_{s=1}^{\nu(n)} \sum_{r=1}^{\nu(n)} A_s(n) A_r(n) a_n^{2s+2r+2} EK^{(2s)}(a_n(x - X)) \times$$
$$\times K^{(2r)}(a_n(x - X)) + \left(1 - \frac{1}{n}\right)(E\hat{g}_n(x))^2. \tag{5.9}$$

Denote

$$E_n(x) = a_n^{2s+2r+2} EK^{(2s)}(a_n(x - X)) K^{(2r)}(a_n(x - X)).$$

It follows from the definition of the kernel $K(x)$ that

$$E_n(x) = a_n^2 \int [a_n^{2s} \sigma^{-2s} \int \phi^{(2s)}(\sigma^{-1}(a_n(x - u) - v) K_1(v)\, dv] \times$$
$$\times [a_n^{2r} \sigma^{-2r} \int \phi^{(2r)}(\sigma^{-1}(a_n(x - u) - t) K_1^{(2r)}(t)\, dt] f(u)\, du.$$

Whence in view of (5.3) and (5.7) we obtain

$$|E_n(x)| = a_n^2 \left| \int \left(\int \sigma^{-1} \phi(\sigma^{-1} a_n(x - u) - v) K_1^{(2s)}(v)\, dv \int \sigma^{-1} \times \right.\right.$$
$$\times \phi(\sigma^{-1} a_n(x - u) - t) K_1^{(2r)}(t)\, dt) f(u) \Big|\, du \le C_2 a_n^2 \times$$
$$\times \frac{\sqrt{2s!}}{s!2^s} \sigma^{-2s} \frac{\sqrt{2r!}}{r!2^r} \sigma^{-2r}. \tag{5.10}$$

Substituting (5.10) into (5.9) we arrive at

$$E\hat{g}_n^2(x) \leq C_3 \frac{a_n^2}{n} (1 - \tau_n)^{-2} + (1 - 1/n)(E\hat{g}_n(x))^2. \tag{5.11}$$

Next, in accordance with (5.9) and (5.11) we have

$$D\hat{g}_n(x) \leq C_3 \frac{a_n^2}{n} (1 - \tau_n)^{-2} + 1/n(Eg_n^{(x)})^2 = O(1). \tag{5.12}$$

Thus (5.8) and (5.12) imply that $E(\hat{g}_n(x) - g(x))^2 \to 0$, q.e.d.

The conditions of Theorem 5.5 are fulfilled if, for example, $\nu(n) = n$, $\tau_n = 1 - (\log n)^{-\beta}$ and $a_n = n^{\alpha}$, $0 < \alpha < 1/2$, $\beta > 0$.

4. Consider yet another method of constructing consistent estimators of the density $g(x)$ based on a wider class of weight functions $K(x)$.

As estimators of $g(x)$ at point x we shall consider a sequence

$$\hat{g}_n(x) = \sum_{s=0}^{\nu(n)} (-1)^s \bar{f}_n^{(2s)}(x) A_s(n),$$

where

$$\bar{f}_n^{(2s)}(x) = (2h_n)^{-2s} \sum_{j=0}^{2s} (-1)^j \binom{2s}{j} f_n(x + (s - j)2h_n),$$

$$f_n(x) = \frac{a_n}{n} \sum_{i=1}^{n} K(a_n(x - X_i)) \quad \text{and} \quad h_n = h(n) \to 0 \quad \text{as} \quad n \to \infty.$$

Here $K(x)$ is a density, $K(x) = K(-x)$ and $x^2K(x) \in L_1(-\infty, \infty)$.

Denote $f_0(x) = Ef_n(x)$. Applying mathematical induction it is easy to verify that

$$E\bar{f}_n^{(2s)}(x) = (2h_n)^{-2s} \int_{x-h_n}^{x+h_n} dt_1 \int_{t_1-h_n}^{t_1+h_n} dt_2 \ldots \int_{t_{2s-1}-h_n}^{t_{2s-1}+h_n} \times$$

$$\times f_0^{(2s)}(t_{2s})\, dt_{2s}.$$

Using this equality we obtain after some simple manipulations that

$$E\hat{g}_n(x) = \sum_{s=0}^{\nu(n)} A_s(n) f^{(2s)}(x) + O(h_n^2(1 - \tau_n)^{-3}) + O(a_n^{-1}(1 - \tau_n))^{-1}. \tag{5.13}$$

Moreover under our assumptions

$$E(\hat{g}_n(x))^2 \leq C_4 \frac{a_n}{n} e^{(\sigma^2\tau_n/h_n^2)} + (1 - 1/n)(E\hat{g}_n(x))^2.$$

Hence

$$D\hat{g}_n(x) = O[(a_n/n) \exp(\sigma^2\tau_n/h_n^2)]. \tag{5.14}$$

From (5.13) and (5.14) and Lemma 5.1 we arrive at

THEOREM 5.6. 1° If $\tau_n^{\nu(n)}(1 - \tau_n)^{-1} \to 0$, $a_n^{-1}(1 - \tau_n)^{-1} \to 0$ and $h_n^2(1 - \tau_n)^{-3} \to 0$ then $E\hat{g}_n(x) \to g(x)$ as $n \to \infty$.

2° If in addition to the conditions stipulated above one requires that

$$\frac{a_n}{n} \exp(\sigma^2 \tau_n^2/h_n^2) \to 0$$

then $\hat{g}_n(x)$ will be a consistent estimator of $g(x)$.

The conditions of the theorem are fulfilled if, for example $\nu(n) = n$,

$$a_n = (\log n)^{\gamma}, \quad h_n = \sigma_0(\log n)^{-1/2}, \quad \tau_n = 1 - (\log n)^{-\alpha}$$

for

$$\sigma_0 > \sigma, \quad 0 < \alpha < \gamma, \quad 0 < \alpha < 1/3.$$

5. We shall now proceed to a solution of the second problem: based on observed values of the sum $X = Y + Z$ estimate the regression function

$$r(x) = E(Y|X = x).$$

It is easy to see that

$$r(x) = \sigma^2 \frac{f^{(1)}(x)}{f(x)} + x,$$

where

$$f(x) = \frac{1}{\sqrt{2\pi}\,\sigma} \int \exp\left(-\frac{(x - u)^2}{2\sigma^2}\right) g(u)\, du.$$

As an estimator of $r(x)$ one could consider the statistic

$$r_n(x) = \sigma^2 \frac{f_n^{(1)}(x)}{f_n(x)} + x.$$

THEOREM 5.7. If $K(x)$ satisfies the conditions of Theorem 5.2 for $s = 1$ and the series $\sum_{n=1}^{\infty} \exp(-\gamma n a_n^{-4}) < \infty$ for any $\gamma > 0$ then $\sup_{x \in I} |r_n(x) - r(x)| \to 0$ a.s., where I is an arbitrary finite interval.

The proof coincides in essense with the proof of Theorem 4.2.1 and it is therefore omitted.

We shall now verify the asymptotic normality of the estimator $r_n(x)$.

Denote $K_1(x) = K^{(1)}(x)$ and $J_0 = \int K_1^2(x)\, dx$.

THEOREM 5.8. If $K(x)$ satisfies the conditions of Theorem 5.1 for $s = 1$, $na_n^{-3} \to \infty$ and $na_n^{-5} \to 0$ as n increases, then the random variable

$$(na_n^{-3})^{1/2}(r_n(x) - r(x))$$

is asymptotically normally distributed with mean zero and variance $\sigma^2(x) = J_0\sigma^2 f^{-1}(x)$.

Proof. Consider

$$f_n^{(1)}(x)f_n^{-1}(x) - Ef_n^{(1)}(x)(Ef_n(x))^{-1} = \xi_n(x)(f_n(x)Ef_n(x))^{-1},$$

where

$$\xi_n = \xi_n(x) = \alpha_n(x)f_n^{(1)}(x) - \beta_n(x)f_n(x),$$

$$\alpha_n = \alpha_n(x) = Ef_n(x); \quad \beta_n = \beta_n(x) = Ef_n^{(1)}(x).$$

Moroever in view of Theorem 5.1 $\lim_{n\to\infty} \alpha_n = f(x)$; $\lim_{n\to\infty} \beta_n = f^{(1)}(x)$.

The variable ξ_n can be rewritten as

$$\xi_n = a_n^2/n \sum_{i=1}^{n} V_i,$$

where

$$V_i = \alpha_n K^{(1)}(a_n(x - X_i)) - \beta_n a_n^{-1} K(a_n(x - X_i)).$$

The mathematical expectation of ξ_n equals zero.

It is easy to verify that

$$na_n^{-3} D(\xi_n) \sim f^3(x)J_0 = \mu(x) = \mu.$$

Whence

$$\lim_{n\to\infty} P\left\{(na_n^{-3})^{1/2}\frac{\xi_n}{\sqrt{\mu}} < \lambda\right\} = \lim_{n\to\infty} P\left\{\frac{\xi_n}{\sqrt{D\xi_n}} < \lambda\right\}. \tag{5.15}$$

We shall show that

$$\lim_{n\to\infty} P\left\{\frac{\xi_n}{\sqrt{D\xi_n}} < \lambda\right\} = \Phi(\lambda). \tag{5.16}$$

To show this it is sufficient to establish that

$$nP\left\{\left|\frac{Z_n}{\sqrt{DZ_n}}\right| \geq \varepsilon n^{1/2}\right\} \to 0 \quad \text{as} \quad n \to \infty \tag{5.17}$$

where

$$Z_n = a_n^2[\alpha_n K^{(1)}(a_n(x - X)) - \beta_n a_n^{-1} K(a_n(x - X))],$$

and in turn to prove (5.17) it is sufficient to verify that

$$\frac{E|Z_n|^{2+\delta}}{n^{\delta/2}(DZ_n)^{1+\delta/2}} \to 0 \quad \text{as} \quad n \to \infty \quad \text{where} \quad \delta > 0. \tag{5.18}$$

Let us prove relation (5.18). We estimate $E|Z_n|^{2+\delta}$:

$$E|Z_n|^{2+\delta} \leq 2^{1+\delta} a_n^{4+2\delta}(|\alpha_n|^{2+\delta} \int K_1^{2+\delta}(a_n(x - u))f(u)\,du +$$

$$+ |\beta_n|^{(2+\delta)} a_n^{-(2+\delta)} \int K^{2+\delta}(a_n(x - u))f(u)\,du) \leq$$

$$\leq C(\delta) a_n^{3+2\delta},$$

where

$$C(\delta) = 2^{1+\delta}\left(\frac{1}{\sqrt{2\Pi}\,\sigma}\right)^{3+\delta}[\int |K_1(x)|^{2+\delta}\,dx + (\int |K_1(x)|\,dx)^{2+\delta} \times$$

$$\times \int |K^{2+\delta}(x)|\,dx].$$

Thus the following bound is valid:

$$\frac{E|Z_n|^{2+\delta}}{n^{\delta/2}(DZ_n)^{1+\delta/2}} \leq \frac{C(\delta)}{(na_n^{-3})^{\delta/2}(a_n^{-3}DZ_n)^{1+\delta/2}\, a_n^{\delta}}.$$

The r.h.s. of the last inequality converges to zero as n increases since $a_n^{-3}DZ_n \to \mu \neq 0$ and $na_n^{-3} \to \infty$.

Taking (5.15) and (5.16) into account we arrive at

$$P\{\sqrt{na_n^{-3}}\,\xi_n(x) < \lambda\mu^{1/2}(x)\} \to \Phi(\lambda). \tag{5.19}$$

Since $f_n(x)Ef_n(x)$ converges to $f^2(x)$ in probability it follows from (5.19) that

$$P((na_n^{-3})^{1/2}\sigma[f_n^{(1)}(x)f_n^{-1}(x) - Ef_n^{(1)}(x)(Ef_n(x))^{-1}] < \lambda\sigma(x)) \to \Phi(\lambda). \tag{5.20}$$

Finally in view of Theorem 5.1 and condition $na_n^{-5} \to 0$ as $n \to \infty$ we obtain from (5.20) that

$$P\{(na_n^{-3})^{1/2}(r_n(x) - r(x)) < \sigma(x)\lambda\} \to \Phi(\lambda), \tag{5.21}$$

q.e.d.

Based on (5.21) one could construct an asymptotic confidence interval for $r(x)$. To accomplish this one should replace the quantity $\sigma^2(x)$ by some consistent estimator choosing for example $\sigma_n^2(x) = J_0\sigma^2 f_n^{-1}(x)$ as such an estimator. Then under the conditions of Theorem 5.8,

$$P\left\{r_n(x) - (J_0 f_n^{-1}(x))^{1/2}\frac{\lambda\sigma^2}{(na_n^{-3})^{1/2}} < r(x) < r_n(x) + \right.$$

$$\left. + (J_0 f_n^{-1}(x))^{1/2}\frac{\lambda\sigma^2}{(na_n^{-3})^{1/2}}\right\} \to \Phi(\lambda).$$

Utilizing the Cramer and Wold result related to multi-variate limit Theorem ([62] p. 86, Section 2b.5) and Rao's theorem ([62], Section 6a.2) one could prove the following general theorem.

THEOREM 5.9. Let $x_1, \ldots, x_n$ be fixed points such that $x_i \neq x_j$, $i \neq j$. Under the conditions of Theorem 5.8 the random vector $(\sqrt{na_n^{-3}}\,(r_n(x_j) - r(x_j)),\ j = \overline{1, m})$ is asymptotically normally distributed with mean 0 and covariance matrix (σ_{ij}), where $\sigma_{ij} = 0$, $i \neq j$, $\sigma_{ii} = \sigma^2(x_i) = J_0\sigma^2 f^{-1}(x_i)$, $i = \overline{1, m}$.

Chapter 5

PROJECTION TYPE NONPARAMETRIC ESTIMATION OF PROBABILITY DENSITY

1. Consistency of Projection-Type Probability Density Estimator in the Norms of Spaces C and L_2

1. As it is known, it was N. N. Chentsov who first proposed a class of nonparametric estimators - referred to as projection estimators - of an unknown probability density $f(x)$ based on independent observations. The idea of his approach is as follows.

Let $X_1, X_2, \ldots, X_n$ be independent random variables whose distribution is given by a probability density $f(x)$, $x \in R = (-\infty, \infty)$. Furthermore, let $L_2(r)$ be the space of functions square integrable with respect to measure μ, $d\mu = r(x)\,dx$, and $\{\phi_j(x)\}$ be an arbitrary orthonormal basis in this space. Assume that the unknown density $f(x) \in L_2(r)$. Denote by $f_N(x)$ the mean square approximation of the density $f(x)$:

$$f_N(x) = \sum_{j=1}^{N} a_j \phi_j(x),$$

where

$$a_j = \int f(x)\phi_j(x)r(x)\,dx = E\alpha_j(X_1), \qquad \alpha_j(x) = \phi_j(x)r(x).$$

Let $F_n(x)$ be the sample distribution function based on a sample of size n. Construct the mean

$$\hat{\alpha}_j = \int \alpha_j(x)\,dF_n(x) = \frac{1}{n}\sum_{i=1}^{n} \alpha_j(X_i)$$

and the polynomial called the projection estimator of $f(x)$ (cf. [6]),

$$\hat{f}_{n,N}(x) = \sum_{j=1}^{N} \hat{\alpha}_j\phi_j(x), \quad N = O(n). \tag{1.1}$$

The projection estimator (1.1) of the density was first introduced and studied by N. N. Chentsov [6]. We present a brief survey of N. N. Chentsov's results as well as of the subsequent authors. Chentsov has shown that by appropriately selecting the number N of terms an interval of a Fourier series one could obtain estimators which converge to $f(x)$ with an optimal rate in the space $L_2(r)$.

Watson [63] generalized the projection estimator (1.1), multiplying $\hat{\alpha}_j$ by an arbitrary factor $\lambda_j(n)$, $j = 1, 2, \ldots$ (multiplier)

$$\hat{f}_{n,N}(x) = \sum_{j=1}^{\nu(n)} \lambda_j(n)\hat{\alpha}_j\phi_j(x), \quad \nu(n) \leq \infty, \tag{1.2}$$

in order to choose in the class (1.2) estimators which are better than any projection estimator in the sense of the minimum of $I_{n,N} = E\|\hat{f}_{n,N} - f\|^2$. Watson obtained an optimal value of the multiplier $\lambda_j(n)$ which yields the minimum of $I_{n,N}$.

In paper [64] the case was studied when a system of Hermite functions serves as a basis of the space L_2 while paper [65] uses Laguerre polynomials for this purpose. Ibragimov and Khas'minskii [66] successfully used Chentsov's method for constructing a nonparametric estimator of a signal S(t) in Gaussian white noise.

2. In this subsection we shall study some asymptotic properties of projection estimators, i.e. convergence to zero of the mean of the squared norm of the error as well as uniform convergence a.s.

THEOREM 1.1. If

$$\lim_{n\to\infty} \int \frac{1}{n}\left(\sum_{i=1}^{N} \alpha_j^2(x)\right) f(x)\, dx = 0,$$

then

$$\lim_{n\to\infty} E\|\hat{f}_{n,N} - f\|^2 = 0.$$

<u>Proof</u>. Note that

$$\|\hat{f}_{n,N} - f\|^2 = \sum_{j=1}^{N} (\hat{\alpha}_j - a_j)^2 + \sum_{j\geq N+1} a_j^2. \tag{1.3}$$

Since $f \in L_2(r)$ the last summand in (1.3) converges to zero as $n \to \infty$. Therefore it is sufficient to show that the mean value of the first sum in (1.3) also converges to zero. Since $E\hat{\alpha}_j = a_j$ and

$$\hat{\alpha}_j - a_j = \frac{1}{n}\sum_{j=1}^{n} (\phi_j(X_i) - a_j),$$

it follows that

$$\sum_{j=1}^{N} E(\hat{\alpha}_j - a_j)^2 = \frac{1}{n}\sum_{i=1}^{N} E\alpha_j^2(X_1) - \frac{1}{n}\sum_{j=1}^{N} a_j^2,$$

and moreover $n^{-1}\sum_{j=1}^{N} a_j^2 \to 0$ as $n \to \infty$, while by the condition the first summand tends to zero. The theorem is thus proved.

THEOREM 1.2. Let r(x) be bounded and let the following conditions be fulfilled:

1° $M = \sup_{j\geq 1} \sup_{x \in R} |\phi_j(x)| < \infty$,

2° the series $\sum_{j=1}^{\infty} a_j \phi_j(x)$ converges to $f(x)$ uniformly,

3° the series $\sum_{n=1}^{\infty} n^{-2} N^5(n)$ is convergent.

Then

$$\sup_{x \in R} |\hat{f}_{n,N}(x) - f(x)| \to 0 \text{ a.s. as } n \to \infty.$$

Proof. We have

$$\sup_{x \in R} |\hat{f}_{n,N}(x) - f(x)| \leq \sup_{x \in R} \left| \sum_{i=1}^{N} (\hat{\alpha}_i - a_i)\phi_i(x) \right| +$$

$$+ \sup_{x \in R} \left| \sum_{i\geq N+1}^{\infty} a_i \phi_i(x) \right|. \tag{1.4}$$

By condition 2° the second command in (1.4) converges to zero while by condition 1°

$$\sup_{x \in R} \left| \sum_{i=1}^{N} (\hat{\alpha}_i - a_i)\phi_i(x) \right| \leq M \sum_{i=1}^{N} |\hat{\alpha}_i - a_i| .$$

We thus obtain

$$P\left\{ \sup_{x \in R} \left| \sum_{i=1}^{N} (\hat{\alpha}_i - a_i)\phi_i(x) \right| \geq \varepsilon \right\} \leq \sum_{i=1}^{N} P\left\{ |\hat{\alpha}_i - a_i| \geq \right.$$

$$\left. \geq \frac{\varepsilon}{MN(n)} \right\} \leq C_1(\varepsilon) N^4(n) \sum_{i=1}^{N} E|\hat{\alpha}_i - a_i|^4. \tag{1.5}$$

The definition of $\hat{\alpha}_i$ yields that

$$E(\hat{\alpha}_i - a_i)^4 = n^{-3} E(\alpha_i(X_1) - a_i)^4 + 3(n^{-2} - n^{-3}) \times$$

$$\times [E(\alpha_i(X_1) - a_i)^2]^2.$$

Whence, since $|\alpha_i(x)| \leq C_1$ and $|a_i| \leq C_2$, we have

$$E(\hat{\alpha}_i - a_i)^4 \leq C_3 n^{-2}.$$

Substituting this inequality into (1.5) we arrive at

$$P\left\{ \sup_{x \in R} \left| \sum_{i=1}^{N} (\hat{\alpha}_i - a_i)\phi_i(x) \right| > \varepsilon \right\} \leq C_2(\varepsilon) n^{-2} N^5(n).$$

Hence in view of conditions 3° applying the Borel-Cantelli lemma we conclude the proof of the theorem.

2. Limiting the Distribution of the Squared Norm of a Projection-Type Density Estimator

1. In this section we consider the problem of the limiting distribution of the squared norm of the error of a projection estimator. The behavior of the goodness of fit test based on this statistic is investigated; namely the asymptotic power under certain local alternatives is calculated. The basic method for obtaining the results is to approximate the normalized and centered sample distribution function by the corresponding Brownian motion processes on an appropriate probability space (cf. Theorem 3.2.1).

Firstly, we shall introduce the following terminology. A random variable x_n which depends on n will be called uniformly bounded if $P\{|x_n| > A\} \to 0$ as $A \to \infty$ uniformly in n. It follows from Chebyshev's inequality that for uniform boundedness of x_n it is sufficient that some positive power of the variable $|x_n|$ possess bounded mathematical expection to n). The following elementary lemma is a corollary of known properties of convergence in probability.

LEMMA 2.1. If a random variable y_n converges in probability to zero and the random variable x_n is uniformly bounded then $z_n = x_n \cdot y_n$ converges in probability to zero as $n \to \infty$.

Let $Y = F(X)$ (here $F(x)$ is a distribution function defined by the density $f(x)$). The variable Y is distributed uniformly on the interval [0, 1]. The collection of random variables $Y_j = F(X_j)$, $j = \overline{1, n}$ has the sample distribution function $S_n(y)$; moreover, it is clear that $S_n(y) = F_n(\psi(y))$, where $\psi(y)$ is the inverse function of $y = F(x)$.

We introduce the empirical process

$$\xi_n(t) = \sqrt{n}\ (S_n(t) - t), \quad 0 \le t \le 1,$$

and the Brownian bridge

$$W_n^0(t) = W_n(t) - tW_n(1), \quad 0 \le t \le 1,$$

where $W_n(t)$ is the standard Wiener process on [0, 1] (cf. Theorem 3.2.1).

Consider sequences of random variables

$$\hat{\xi}_j = \sqrt{n}\ (\hat{\alpha}_j - a_j) = \int \alpha_j(x)\ d\xi_n(F(x)),$$

$$\xi_j = \int \alpha_j(x)\ dW_n^0(F(x)),$$

$$w_j = \int \alpha_j(x)\ dW_n(F(x)), \quad j = 1, 2, \ldots, N$$

and set

$$T_{n,N}^{(1)} = \frac{1}{N} \sum_{j=1}^{N} \xi_j^2, \quad T_{n,N}^{(2)} = \frac{1}{N} \sum_{j=1}^{N} w_j^2. \tag{2.1}$$

We shall investigate the limiting distribution of the variable (as $n, N \to \infty$)

$$T_{n,N} = \frac{n}{N} \| \hat{f}_{n,N} - E\hat{f}_{n,N} \|^2. \tag{2.2}$$

It is easy to verify that

$$T_{n,N} = \frac{1}{N} \sum_{j=1}^{N} \hat{\xi}_j^2$$

Assume that $f(x)r(x)$ is bounded and that $\alpha_j(x)$, $j = 1, 2, \ldots$, are each of bounded variation $V_j < \infty$. Set

$$b_N = \sum_{j=1}^{N} V_j \quad \text{and} \quad d_{n,N} = b_N n^{-1/2} \log n.$$

LEMMA 2.2. If $d_{n,N} = O(\sqrt{N})$, then

1° $\sqrt{N}\,(T_{n,N} - T_{n,N}^{(1)})$ converges to zero in probability as $n, N \to \infty$;

2° $\sqrt{N}\,(T_{n,N}^{(1)} - T_{n,N}^{(2)})$ converges to zero in probability as $n, N \to \infty$;

Proof. Applying the formula of integration by parts we arrive at

$$|\hat{\xi}_j - \xi_j| \leq 2 \sup_{0 \leq t \leq 1} |\xi_n(t) - W_n^0(t)| \cdot V_j.$$

Whence, introducing the notation

$$\ell_j(n) = |\hat{\xi}_j| + |\xi_j|, \quad A_{n,N}^{(1)} = \frac{2b_N}{\sqrt{N}} \sup_{0 \leq t \leq 1} |\hat{\xi}_n(t) - W_n^0(t)|$$

and

$$A_{n,N}^{(2)} = b_N^{-1} \sum_{j=1}^{N} V_j \ell_j(n),$$

we obtain

$$\sqrt{N}\, |T_{n,N} - T_{n,N}^{(1)}| \leq A_{n,N}^{(1)} A_{n,N}^{(2)}.$$

In accordance with Theorem 3.2.1 the random variable $A_{n,N}^{(1)} \sqrt{N}\, d_{n,N}^{-1}$ is uniformly bounded. We shall prove that the random variable $A_{n,N}^{(2)}$ is also uniformly bounded. To prove this it is sufficient to verify that $E\ell_j(n) < C_1 < \infty$, $j = 1, 2, \ldots$. The inequality $E\ell_j(n) < C_1$ is a corollary of the estimate

$$E|\hat{\xi}_j| \leq \sqrt{n}\ \{D\hat{\alpha}_j\}^{1/2} = (D\alpha_j(X_1))^{1/2} \leq (\int \alpha_j^2(x) f(x)\, dx)^{1/2} \leq$$
$$\leq C_2 < \infty$$

and

$$E|\xi_j| \le E\left|\int \alpha_j(x)\ dW_n(F(x))\right| + E|W_n(1)|\left|\int \alpha_j(x)f(x)\ dx\right| \le$$

$$\le \left(\int \alpha_j^2(x)f(x)\ dx\right)^{1/2} + E|W_n(1)|\left(\int f^2(x)r(x)\ dx\right)^{1/2} \le$$

$$\le C_3 < \infty.$$

Hence in view of Lemma 2.1 assertion 1° is valid.

We now prove 2°. From (2.1) we have

$$\sqrt{N}\,|T_{n,N}^{(1)} - T_{n,N}^{(2)}| \le N^{-1/2}\left\{[W_n(1)]^2 \sum_{j=1}^{N}\left(\int \alpha_j(x)f(x)\ dx\right)^2 + \right.$$

$$\left. + 2|W_n(1)|\left|\int\int \sum_{j=1}^{N} \alpha_j(u)\alpha_j(x)f(u)\ dW_n(F(x))\ du\right|\right\} =$$

$$= \frac{1}{\sqrt{N}}\ A_{n,N}^{(3)}.$$

However as $N \to \infty$

$$\sum_{j=1}^{N}\left(\int \alpha_j(x)f(x)\ dx\right)^2 = \sum_{j=1}^{N} \alpha_j^2 \to \int f^2(x)r(x)\ dx. \tag{2.3}$$

and

$$E\left\{\int\int \sum_{j=1}^{N} \alpha_j(x)\alpha_j(u)f(x)\ du\ dW_n(F(x))\right\}^2 =$$

$$= E\left(\int\left[\sum_{j=1}^{N} a_j\alpha_j(x)\right] dW_n(F(x))\right)^2 =$$

$$= \int\left[\sum_{j=1}^{N} a_j\alpha_j(x)\right]^2 f(x)\ dx \le C_4\,\|f_N\|^2, \tag{2.4}$$

moreover $\|f_N\|^2 \to \|f\|^2$ as $n \to \infty$. In view of (2.3) and (2.4) we conclude that $A_{n,N}^{(3)}$ is uniformly bounded. Thus assertion 2° is verified and with it the lemma is proved.

It follows from Lemma 2.2 that $\sqrt{N}\ (T_{n,N} - T_{n,N}^{(2)}) \to 0$ in probability as $n,\ N \to \infty$.

The random vector $(W_1, W_2, \ldots, W_N)$ (cf. formula (2.1)) possesses a multivariate normal distribution with zero mean vector and covariance matrix $C = (a_{ij})$ where

$$a_{ij} = \int \alpha_i(x)\alpha_j(x)f(x)\ dx. \tag{2.5}$$

Consider now the characteristic function $\phi_{n,N}(t)$ of the variable $T_{n,N}^{(2)}$ (2.1). It is defined by the equality

$$\phi_{n,N}(t) = |I - 2itCD|^{-1/2} = \prod_{j=1}^{N} (1 - 2\lambda_j it)^{-1/2} =$$

$$= \exp\left\{\sum_{m=1}^{\infty} 2^{m-1}(it)^m S_m(N) \frac{1}{m}\right\} \quad \text{for} \quad |t| \leq \frac{1}{2} S_p(CD), \tag{2.6}$$

where $\lambda_1, \lambda_2, \ldots, \lambda_N$ are the roots of the equation $|\vec{C}D - \lambda I| = 0$, D is a diagonal matrix with elements 1/N and $S_m(N) = \sum_{j=1}^{N} \lambda_j^m = S_p(CD)^m$. It is not too difficult to verify that

$$S_m(N) = N^{-m} \sum_{i_1=1}^{N} \cdots \sum_{i_m=1}^{N} a_{i_1 i_2} \cdot a_{i_2 i_3} \cdots a_{i_m i_1}. \tag{2.7}$$

Introduce the notation

$$K_N(x, y) = \sum_{j=1}^{N} \phi_j(x)\phi_j(y)r(y),$$

$$\Delta_N = \frac{1}{N} \sum_{j=1}^{N} \int \alpha_j^2(x) f(x)\, dx, \tag{2.8}$$

$$\sigma_N^2 = \frac{2}{N} \sum_{i=1}^{N} \sum_{j=1}^{N} \left(\int \alpha_i(x)\alpha_j(x) f(x)\, dx\right)^2.$$

Utilizing (2.5) and (2.8) we obtain from (2.7) that

$$S_m(N) = N^{-m} \int \cdots \int K_N(x_1, x_2)K_N(x_2, x_3) \cdots K_N(x_m, x_1) \times$$

$$\times f(x_1) \cdots f(x_m) r(x_1) \cdots r(x_m)\, dx_1 \cdots dx_m.$$

THEOREM 2.1. If

$$d_{n,N} = O(N^{1/2}), \quad \Delta_N = \Delta + O(N^{-1/2}), \quad \sigma_N^2 = \sigma^2 + O(N^{-1/2})$$

as $N \to \infty$ and for each $m \geq 3$

$$Q_m(N) \equiv N^{m-1} S_m(N) = O(1), \quad \text{as} \quad N \to \infty, \tag{2.9}$$

then as $n, N \to \infty$ the distribution of the random variable $U_{n,N} = \sqrt{N}(T_{n,N} - \Delta)$ converges to the normal distribution with mean 0 and variance σ^2.

Proof. As Lemma 2.2 implies, to prove the theorem it is sufficient to obtain the limiting distribution of the variable

$$U^*_{n,N} = \sqrt{N}\, (T^{(2)}_{n,N} - \Delta_N)\sigma_N^{-1}.$$

Denote the characteristic function of $U^*_{n,N}$ by $\phi^*_{n,N}(t)$. We have from (2.6) that

$$\ln\phi^*_{n,N}(t) = -\frac{\Delta_N it}{\sigma_N/\sqrt{N}} + \sum_{m=1}^{\infty} \frac{2^{m-1}}{m} \left(\frac{it}{\sigma_N/\sqrt{N}}\right)^m S_m(N). \qquad (2.10)$$

Taking into account that $S_1(N) = \Delta_N$ and $\sigma_N^2 = 2NS_2(N)$ we obtain from (2.10) the cumulants of the random variable $U^*_{n,N}$: $\chi_1 = 0$, $\chi_2 = 1$,

$$\chi_m = 2^{m-1}(m-1)! \frac{N^{m/2}}{\sigma_N^m} S_m(N) = 2^{m-1}(m-1)! \frac{Q_m(N)}{\sigma_N^m} \frac{1}{N^{(m/2)-1}} =$$

$$= O\left(\frac{1}{N^{(m/2)-1}}\right), \quad m \geq 3. \qquad (2.11)$$

Relation (2.11) indicates that the moments of the random variable $U^*_{n,N}$ converge as n and $N \to \infty$ to the moments of the standard normal distribution. The theorem is thus proved.

Denote

$$\tilde{T}^2_{n,N} = \frac{n}{N} \|\hat{f}_{n,N} - f\|^2 \quad \text{and} \quad \delta_{n,N} = \frac{n}{N} \|f_N - f\|^2.$$

From Theorem 2.1 follows

COROLLARY. If the conditions of Theorem 2.1 are fulfilled and $\sqrt{N}\delta_{n,N} \to 0$ then $\sqrt{N}(\tilde{T}_{n,N} - \Delta)$ is distributed asymptotically normally with parameters $(0, \sigma)$.

The proof follows from the decomposition $\tilde{T}_{n,N} = \delta_{n,N} + T_{n,N}$.

If for an unknown density $f(x)$ conditions

$$|a_k| = |\int \phi_k(x) f(x) r(x)\, dx| \leq C_5 k^{-m}, \quad k = 1, 2, \ldots,$$

are a priori fulfilled then for the choice of $N = n^\alpha$, $\alpha > 2/(4m-1)$, the condition $\sqrt{N}\,\delta_{n,N} \to 0$ is fulfilled. Indeed, from Parseval's inequality

$$\| f_N - f \|^2 = \sum_{k=N+1}^{\infty} a_k^2 \leq C_6(m) N^{-2m+1},$$

whence

$$\sqrt{N}\,\delta_{n,N} = \frac{n}{\sqrt{N}} \|f_N - f\|^2 \leq C_6(m) n^{1-\alpha(2m-\frac{1}{2})} = o(1).$$

2. Consider several examples of the application of Theorem 2.1.

a. Let X be a bounded random variable $C' \leq X \leq C''$, $\phi_j(x) = \sqrt{N(C'' - C')^{-1}}\, \chi_j(x)$ (cf. [6]), where $\chi_j(x)$ is the indicator function of the interval $C' + (j-1)\omega \leq x \leq C' + j\omega$, $\omega = (C'' - C')N^{-1}$. In this case $r(x) \equiv 1$. Condition $d_{n,N} = O(\sqrt{N})$ is satisfied for $N = n^\alpha$, $\alpha < 1/2$, since

$b_N = 2(C'' - C')N^{3/2}$. The remaining conditions are also fulfilled provided $f'(x)$ is bounded. It is easy to calculate that

$$\sigma_N^2 = \frac{2}{C'' - C'} \int_{C'}^{C''} f^2(x)\,dx + O(N^{-1}),$$

$$Q_m(N) = \int_{C'}^{C''} [f(x)]^m\,dx + O\left(\frac{1}{N}\right).$$

b. Let $-\pi \leq X \leq \pi$ and $\phi_j(x)$, $j = 1, 2, \ldots$, be a system of trigonometric functions on $[-\pi, \pi]$.

It is easy to observe that $b_N = O(N^2)$, $d_{n,N} = O(\sqrt{N})$ for $N = n^{\alpha}$, $\alpha < 1/3$ and that in (2.2) n/N should be replaced by $n/(2N + 1)$. As it is known that in the case under consideration $K_N(x, y)$ is the Dirichlet kernel. Taking this into account one can represent σ_N^2 by means of the so-called Fejér integral:

$$\sigma_N^2 = \frac{1}{\pi} \int \left[\frac{1}{(2N + 1)\pi} \int \frac{\sin^2(2N + 1)u}{\sin^2 u} f(x_2 + 2u)\,du\right] f(x_2)\,dx_2 \tag{2.12}$$

Next, assuming that $f'(x)$ is bounded it follows from Theorem 3.15 in [67] that the inner integral in (2.12) equals $f(x_2) + O(N^{-1} \ln N)$ and hence

$$\sigma_N^2 = \frac{1}{\pi} \int f^2(x)\,dx + O\left(\frac{1}{\sqrt{N}}\right). \tag{2.13}$$

It follows directly from the definition of Δ_n that

$$\Delta_N = \frac{1}{2\pi} + O(N^{-1/2}). \tag{2.14}$$

Condition (2.9) is also fulfilled. Indeed,

$$|Q_m(N)| \leq C_7 N^{-1}(L_N)^m, \quad m \geq 3, \tag{2.15}$$

where $L_N \sim 4\pi^{-2} \ln N$ is the Lebesgue constant[67]. Thus, taking (2.13), (2.14) and (2.15) into account we conclude from Theorem 2.1 that the random variable

$$(2N + 1)^{1/2}\left(T_{n,N} - \frac{1}{2\pi}\right)\left(\frac{1}{\pi} \int f^2(x)\,dx\right)^{-1/2}$$

is distributed asymptotically normally with parameters (0, 1). From here it is possible to obtain more refined results for asymptotic confidence bounds than those given by Kronmal and Tarter in [68].

c. Let

$$\phi_j(x) = \sqrt{\frac{2j + 1}{2}}\, P_j(x), \quad x \in [-1, 1], \quad j = 1, 2, \ldots,$$

where $P_j(x)$ are Legendre polynomials. In this case $r(x)\,dx = dx$.

Let $d_0(x)$ be a positive and bounded function on $[-1, 1]$. Set $d(x) = d_0(x) I_{[-1,1]}(x)$, where $I_A(x)$ is the indicator function of the set A. It is easy to verify that the projection estimator for the function $p(x) = f(x)\,d(x)$ is of the form

$$\hat{p}_{n,N}(x) = \sum_{j=1}^{N} \hat{\alpha}_j \phi_j(x), \qquad \hat{\alpha}_j = \frac{1}{n} \sum_{i=1}^{n} d(X_i)\phi_j(X_i),$$

and for $f(x)$ it is

$$\hat{f}_{n,N}(x) = \hat{p}_{n,N}(x)/d(x).$$

Our aim is to obtain the limiting distribution of the functional

$$T_{n,N} = \frac{n}{N} \int [\hat{p}_{n,N}(x) - E\hat{p}_{n,N}(x)]^2\, dx =$$

$$= \frac{n}{N} \int [\hat{f}_{n,N}(x) - E\hat{f}_{n,N}(x)]^2\, d^2(x)\, dx.$$

Assume that $f(x)$ and $d_0(x)$ possess bounded derivatives on $[-1, 1]$ and $d_0(x) = O((1 - x^2)^\gamma)$, $\gamma \geq 3/4$, as $x \uparrow 1$ or $x \downarrow -1$.

Utilizing the inequality $\int_{-1}^{1} |P_j(x)|\, dx \leq 2/(\sqrt{2j+1})$ from the relation

$$\frac{dP_m(x)}{dx} = \sum_{k=1}^{s} (2m - 4k + 3) P_{m-2k+1}(x),$$

where $s = m/2$ (m even) and $s = (m+1)/2$ (m odd) we obtain that $\overset{1}{\underset{-1}{V}}(\phi_j d_0) \leq C_7 j^2$. Hence $b_N = O(N^3)$ and $d_{n,N} = O(\sqrt{N})$ for $N = n^\alpha$, $0 < \alpha < 1/5$.

In (2.5) one should set $f(x)d^2(x)$ in place of $f(x)$. Let $\lambda_1^{(N)}, \lambda_2^{(N)}, \ldots$, be the eigen values of a Toeplitz form with the matrix

$$M_N(f) = \{ \int \phi_j(x)\phi_i(x) f(x)\, d^2(x)\, dx, \quad j = 1, 2, \ldots, N\}.$$

Then

$$Q_m(N) = N^{m-1} s_m(N) = \frac{1}{N} \sum_{i=1}^{N} (\lambda_i^{(N)})^m.$$

Therefore from formula (4) on page 89 of [69] (Chapter 6, Section 3), we have

$$Q_m(N) \to \frac{1}{\pi} \int_{-1}^{1} f^m(x) d_0^{2m}(x) \frac{1}{\sqrt{1 - x^2}}\, dx \quad \text{as} \quad N \to \infty. \tag{2.16}$$

Consequently $Q_m(N) = O(1)$ for each fixed $m \geq 3$. Furthermore, since $\sigma_N^2 = 2NS_2(N) = 2Q_2(N)$ we obtain from (2.16) that

$$\sigma_N^2 \to \frac{2}{\pi} \int_{-1}^{1} f^2(x)\, d_0^4(x) \frac{1}{\sqrt{1-x^2}}\, dx. \tag{2.17}$$

It remains only to investigate the asymptotics of Δ_N as $N \to \infty$.

For the Legendre polynomials with $0 < \Theta < \pi$ the following asymptotic formula is valid (see, e.g. [70]):

$$P_s(\cos\Theta) = \lambda_s \frac{\cos\left[(s + 1/2)\Theta - \frac{\pi}{4}\right]}{(\sin\Theta)^{1/2}} + \frac{R_s(\Theta)}{(\sin\Theta)^{1/2}}, \tag{2.18}$$

In this formula the remainder is bounded by

$$|R_s(\Theta)| \leq \frac{C_8}{s^{3/2}\Theta}\left(\frac{\pi}{2} - \Theta\right), \quad 0 < \Theta \leq \frac{\pi}{2}, \tag{2.19}$$

$$|R_s(\Theta)| \leq \frac{C_8}{s^{3/2}(\pi - \Theta)}\left(\Theta - \frac{\pi}{2}\right), \quad \frac{\pi}{2} \leq \Theta < \pi, \tag{2.20}$$

where the constant C_8 does not depend on s and Θ and the coefficient λ_s admits the representation $\lambda_s = \sqrt{2/\pi \cdot s} + O(s^{-3/2})$.

Utilizing (2.18) we obtain that

$$\int_{-1}^{1} \phi_j^2(x)\, d_0^2(x) f(x)\, dx = \int_0^{\pi} \phi_j^2(\cos\Theta)\, d_0^2(\cos\Theta) f(\cos\Theta) \times$$

$$\times \sin\Theta\, d\Theta = I_1 + I_2 + I_3,$$

where

$$I_1 = \frac{2j+1}{2} \lambda_j^2 \int_0^{\pi} \cos^2\left[\left(j + \frac{1}{2}\right)\Theta - \frac{\pi}{4}\right] d_0^2(\cos\Theta) f(\cos\Theta)\, d\Theta,$$

$$I_2 = (2j+1)\lambda_j \int_0^{\pi} \cos\left[\left(j + \frac{1}{2}\right)\Theta - \frac{\pi}{4}\right] R_j(\Theta)\, d_0^2(\cos\Theta) \times$$

$$\times f(\cos\Theta)\, d\Theta,$$

$$I_3 = \frac{2j+1}{2} \int_0^{\pi} R_j^2(\Theta)\, d_0^2(\cos\Theta) f(\cos\Theta)\, d\Theta.$$

It is easy to verify that

$$I_1 = \frac{1}{\pi} \int_0^{\pi} d_0^2(\cos \Theta) f(\cos \Theta)\, d\Theta \left(1 + \frac{1}{j}\right) + \frac{2j + 1}{4} \lambda_j^2 \times$$

$$\times \int_0^{\pi} \sin(2j + 1)\Theta\, d_0^2(\cos \Theta) f(\cos \Theta)\, d\Theta, \tag{2.21}$$

and hence applying integration by parts we check that the last term in (2.21) is $O(1/j)$.

Whence

$$I_1 = \frac{1}{\pi} \int_{-1}^{1} f(x)\, d_0^2(x) \frac{1}{\sqrt{1 - x^2}}\, dx + O\left(\frac{1}{j}\right). \tag{2.22}$$

In view of the bounds (2.19) and (2.20) we obtain

$$I_2 = O\left(\frac{1}{j}\right) \int_{-1}^{1} \frac{f(x)\, d_0^2(x)\, dx}{1 - x^2}, \tag{2.23}$$

$$I_3 = O\left(\frac{1}{j^2}\right) \int_{-1}^{1} \frac{f(x)\, d_0^4(x)}{(1 - x^2)^{3/2}}\, dx \tag{2.24}$$

The integral appearing in (2.23) and (2.24) are finite since $d_0(x) = O((1 - x^2)^{\gamma})$, $\gamma \geq 3/4$, as $x \uparrow 1$ or $x \downarrow -1$. It follows from (2.22), (2.23) and (2.24) that

$$\int_{-1}^{1} \phi_j^2(x)\, d_0^2(x) f(x)\, dx = \frac{1}{\pi} \int_{-1}^{1} f(x)\, d_0^2(x) \frac{1}{\sqrt{1 - x^2}}\, dx + O\left(\frac{1}{j}\right).$$

Therefore

$$\Delta_N = \frac{1}{\pi} \int_{-1}^{1} f(x)\, d_0^2(x) \frac{dx}{\sqrt{1 - x^2}} + O\left(\frac{\log N}{N}\right). \tag{2.25}$$

Thus taking (2.16), (2.17) and (2.25) into account we obtain from Theorem 2.1 that the random variable

$$N^{1/2} \left[T_{n,N} - \frac{1}{\pi} \int_{-1}^{1} f(x)\, d_0^2(x) \frac{1}{\sqrt{1 - x^2}}\, dx\right]$$

is asymptotically normally distributed with parameters $(0, \sigma)$, where

$$\sigma^2 = \frac{2}{\pi} \int_{-1}^{1} f^2(x)\, d_0^4(x) \frac{1}{\sqrt{1 - x^2}}\, dx.$$

3. Let it be required to test the hypothesis that the probability density of X_j coincides with $f_0(x)$, i.e. $H_0{:}f(x) = f_0(x)$ where $f_0(x)$ is a given density and furthermore, let H be a family of alterantives close to H_0 of the form:

$$H : f(x) = f_0(x) + \gamma_n \phi(x), \quad \gamma_n \downarrow 0 \quad \text{as} \quad n \to \infty. \tag{2.26}$$

One can test H_0 versus the alternatives (2.26) by using for this purpose statistic $T_{n,N}$ and choosing the critical region of the form $\{T_{n,N} \geq t_\alpha\}$ where, in view of Theorem 2.1 we have approximately $t_\alpha = \Delta_0 + N^{-1/2}\lambda_\alpha \sigma_0$ denotes the $(1 - \alpha)$ quantile of the standard normal distribution, i.e. $\Phi(\lambda_\alpha) = 1 - \alpha$, Δ_0 and σ_0 are parameters appearing in Theorem 2.1 calculated under H_0. It turns out that for these alternatives a test based on $T_{n,N}$ possesses a non-trivial limiting power. More precisely, the following theorem is valid.

THEOREM 2.2. Let $f(x) \in L_2(r)$ and satisfy the conditions of Theorem 2.1 and let the series of Fourier coefficients of the function $\phi(x)$ be absolutely convergent. Then if there exists an $0 < \alpha_0 < 2/3$ such that $d_{n,N} = O(\sqrt{N})$ for $N = n^\alpha$, $0 < \alpha \leq \alpha_0$ and $\gamma_n = n^{(\alpha-2)/4}$ as n and $N \to \infty$, the probability

$$P_H\left\{ T_{n,N} \geq \Delta_0 + \frac{\lambda_\alpha}{\sqrt{N}} \sigma_0 \right\} \to 1 - \phi(\lambda_\alpha - \sigma_0^{-1} \int \phi^2(x) r(x)\, dx). \tag{2.27}$$

<u>Proof</u>. Denote by $\tilde{\Delta}_N$ and $\tilde{\sigma}_N^2$ the values of Δ_N and σ_N^2 under H and set

$$E(\hat{f}_{n,N}(x)/H_0) = f_N^0(x), \qquad E(\hat{f}_{n,N}(x)/H) = f_N(x),$$

$$\frac{n}{N} || \hat{f}_{n,N} - f_N ||^2 = \tilde{T}_n.$$

Clearly, $\tilde{\sigma}_N^2 = \sigma_0^2 + O(\gamma_n) + O(N^{-1/2})$ and moreover $\sqrt{N}\,\gamma_n \to 0$ as $n \to \infty$. We have

$$\sqrt{N} \left(\frac{T_{n,N} - \Delta_0}{\sigma_0} \right) = \sqrt{N} \left(\frac{\tilde{T}_n - \tilde{\Delta}_N}{\sigma_0} \right) + 2 \frac{n}{\sqrt{N}} \frac{1}{\sigma_0} \int (\hat{f}_{n,N}(x) -$$

$$- f_N(x))(f_N(x) - f_N^0(x)) r(x)\, dx + \frac{n}{\sigma_0 \sqrt{N}} \times$$

$$\times || f_N - f_N^0 ||^2 + \sqrt{N} \frac{(\tilde{\Delta}_N - \Delta_0)}{\sigma_0} =$$

$$= U_1 + U_2 + U_3 + U_4. \tag{2.28}$$

In accordance with Theorem 2.1 the random variable U_1 under H is distributed asymptotically normally with parameters (0, 1).

Set $\phi_N^0(x) = \sum_{j=1}^{N} \phi_j \phi_j(x)$ where ϕ_j are the Fourier coefficients of the function $\phi(x)$. Then

$$f_N(x) = f_N^0(x) + \gamma_n \phi_N^0(x).$$

It follows directly from relation (2.29) that

$$U_3 = \frac{1}{\sigma_0}\frac{n}{\sqrt{N}}\,||f_N - f_N^0||^2 = \frac{1}{\sigma_0}\,||\phi_N^0||^2 \to \frac{||\phi||^2}{\sigma_0} \quad \text{as} \quad N \to \infty.$$

Furthermore,

$$U_4 = n^{(3a-2)/4}\frac{1}{\sigma_0 N}\sum_{j=1}^{N}\int \alpha_j^2(x)\phi(x)\,dx \to 0 \quad \text{as} \quad n \to \infty.$$

Consider now U_2. Substituting the expression for $f_N(x)$ from (2.29) into U_2 we obtain

$$U_2 = \frac{2n}{\sqrt{N}}\frac{\gamma_n}{\sigma_0}\sum_{j=1}^{N}(\hat{\alpha}_j - a_j)\phi_j = \frac{2n^{1/2}}{\sqrt{N}}\gamma_n\frac{1}{\sigma_0}\,z_n,$$

where

$$z_n = \sqrt{N}\sum_{j=1}^{N}(\hat{\alpha}_j - a_j)\phi_j \quad \text{and} \quad a_j = E(\phi_j(X_1)r(X_1/H)).$$

Using the same arguments as in the proof of the Lemma 2.2 it can be shown that

$$z_n - \sum_{j=1}^{N}\tilde{W}_j\phi_j \to 0$$

in probability as $n \to \infty$, where $\tilde{W}_j$ is defined under H in the same manner as W_j in formula (2.1).

The random variable

$$R_n = \left(\sum_{j=1}^{N}\tilde{W}_j\phi_j\right)D_n^{-1/2}, \qquad D_n = \sum_{i=1}^{N}\sum_{j=1}^{N}\tilde{a}_{ij}\phi_i\phi_j > 0$$

$$\tilde{a}_{ij} = E(\tilde{W}_i\tilde{W}_j),$$

is distributed normally with parameters (0, 1) where $\tilde{a}_{ij} = a_{ij}^0 + O(\gamma_n)$. From here and from the bounds $D_n \le C_8(\sum_{j=1}^{N}|\phi_j|)^2 \le C_9$ it follows that $D_n^{1/2}(n^{1/2}\gamma_n)/(\sqrt{N}\sigma_0)R_n \to 0$ in probability as $n \to \infty$. Hence $U_2 \to 0$ in probability as $n \to \infty$. The theorem is proved.

Now let the hypothesis H_0 stipulating the uniform distribution on [0, 1] be tested. For each n the alternative H is given by the density

$$f(x) = 1 + \gamma_n\phi(x) + O(\gamma_n), \qquad \gamma_n = n^{-(1/2)+\alpha/4}, \quad 0 \le x \le 1, \tag{2.30}$$

where $O(\gamma_n)$ is uniformly in x and $\phi(x)$ is a fixed function such that $\int \phi(x)\,dx = 0$ and $\phi(x) \in L_2$.

Let, as in example a) $\phi_j(x) = \sqrt{N}\,\chi_j(x)$ where $\chi_j(x)$ is the indicator of the interval $I(j) = \{x: (j - 1)\omega \leq x \leq j\omega\}$; $\omega = 1/N$ and let m_j be the number of sample points on the interval $I(j)$. For testing H_0 versus H the statistic

$$\chi_n^2 = \sum_{j=1}^{N} nN \left(\frac{m_j}{n} - 1/N \right)^2$$

is used. The results obtained in [71] imply that for equiprobable groups the power of the χ_n^2-test relative to each 'close' alternative H_1 of the form $H_1 = \{f(x) = 1 + n^{-1/2}\phi(x) + o(n^{-1/2})\}$ converges to the level of the test as the number of the groups N increases (the limiting χ_n^2 distributions under H_0 and H_1 become indistinguishable as N increases with n). However, in view of Theorem 2.2 it turns out that for the alternatives (2.30) the χ_n^2-test possesses a nontrivial limiting power. Indeed, as it is easy to calculate, $\Delta_0 = 1$, $\sigma_0^2 = 2$, $\chi_n^2/N = T_{n,N}$ and hence

$$\frac{\chi_n^2 - N}{\sqrt{2N}} = \sqrt{N} \left(\frac{T_{n,N} - \Delta_0}{\sigma_0} \right) .$$

From here in view of Theorem 2.1 it follows that $(2N)^{-1/2}(\chi_n^2 - N)$ is distributed asymptotically normally (0, 1) (cf. also [54]). The critical region is $\{\chi_n^2 \geq N + \lambda_\alpha \sqrt{2N}\}$ and the power β_N of the test relative to the alternative (2.30) in view of Theorem 2.2 is approximately equal to

$$\beta_N \cong 1 - \phi \left(\lambda_\alpha - \frac{1}{\sqrt{2}} \int \phi^2(x)\, dx \right) .$$

Remark. One can also investigate the asymptotics of the power of the $T_{n,N}$-test relative to each 'singular' close alternative introduced in Section 3 of Chapter 3. These investigations essentially coincide with investigations presented in Section 3 of Chapter 3 and are therefore omitted.

Some Examples of the Kernel K(x)

K(x)	$\int K^2(x)\, dx$
$\frac{1}{2}$, $\lvert x\rvert \leq 1$ 0, $\lvert x\rvert > 1$	$\frac{1}{2}$
$1 - \lvert x\rvert$, $\lvert x\rvert \leq 1$ 0, $\lvert x\rvert > 1$	$\frac{2}{3}$
$\frac{4}{3} - 8x^2 + 8\lvert x\rvert^3$, $\lvert x\rvert < \frac{1}{2}$ $\frac{8}{3}(1 - \lvert x\rvert)^3$, $\frac{1}{2} \leq \lvert x\rvert \leq 1$ 0, $\lvert x\rvert > 1$	0.96
$\frac{1}{\sqrt{2\pi}} \exp\left\{-\frac{x^2}{2}\right\}$	$\frac{1}{2\sqrt{\pi}}$
$\frac{1}{2} \exp\{-\lvert x\rvert\}$	$\frac{1}{4}$
$\frac{1}{\pi}\left(\frac{\sin x}{x}\right)^2$	$\frac{2}{3\pi}$
$\frac{1}{\sqrt{6}} - \lvert x\rvert/6$, $\lvert x\rvert \leq \sqrt{6}$ 0, $\lvert x\rvert > \sqrt{6}$	$\frac{\sqrt{6}}{9}$
$\frac{3}{2}\left(1 - \frac{x^2}{3}\right) \frac{1}{\sqrt{2\pi}} \exp\left\{-\frac{x^2}{2}\right\}$	$\frac{27}{32\sqrt{\pi}}$
$\frac{15}{8}\left(1 - \frac{2}{3}x^2 + \frac{1}{15}x^4\right) \frac{1}{\sqrt{2\pi}} \exp\left\{-\frac{x^2}{2}\right\}$	$\frac{2625}{2048\sqrt{\pi}}$
$\frac{3}{4\sqrt{5}} - \frac{3x^2}{20\sqrt{5}}$, $\lvert x\rvert \leq \sqrt{5}$ 0, $\lvert x\rvert > \sqrt{5}$	$\frac{3}{5\sqrt{5}}$
$0.54 + 0.46 \cos \pi x$, $\lvert x\rvert \leq 1$ 0, $\lvert x\rvert > 1$	$\frac{3}{4}$
$\frac{9}{8}\left(1 - \frac{5}{3}x^2\right)$, $\lvert x\rvert \leq 1$ 0, $\lvert x\rvert > 1$	$\frac{9}{8}$

Addendum*

LIMITING DISTRIBUTION OF QUADRATIC DEVIATION FOR A WIDE CLASS OF PROBABILITY DENSITY ESTIMATORS

As it was mentioned in Chapter 3, Section 2, the limiting distribution of the quadratic deviation $\int (f_n(x) - f(x))^2 r(x)\, dx$ of estimators $f_n(x)$ of the density $f(x)$ constructed by means of 'kernels' was studied by P. Bickel and M. Rosenblatt [49]. An analogous problem for estimators constructed by means of systems of orthonormal functions ('projection estimators') was investigated by E. A. Nadaraya [91]. The basic method utilized in these investigations was the Breiman-Brillinger approximation [52].

An additional randomization - when, instead of n observations a random number of observations distributed according to the Poisson distribution with the mean value n which is independent of observations - constitutes the second approach to investigation of the limiting distribution of the quadratic deviation ([56], [82]).

Problems of the limiting distribution of quadratic deviations of density estimators of both types were the subject of subsequent investigations. E. A. Nadaraya [100] and P. Hall [101] simultaneously proposed a martingale method for investigation of the limiting distribution of quadratic deviations of kernel-type estimators while for projection estimators a similar method was considered by J. Ghorai [102]. In [100] the central limit theorem for the Liptser-Shiryaev [103] semimartingales is utilized. Application of this method allowed us to relax a number of conditions as compared with the previously available in the literature ([49], [91]).

The purpose of this addendum is to obtain the limiting distribution of the quadratic deviation for density estimators $f_n(x)$ of a general type, including kernel-type estimators, projection estimators, histograms and so on. The proposed method allows us, using a unified approach, to study the limiting distribution of quadratic deviations in one-dimensional as well as multidimensional cases. The proof of the basic Theorem 1 is carried out along the lines of the proof of Theorem 1 presented in [100].

Let $X_1, X_2, \ldots, X_n$ be independent, identically distributed random variables (one- or multidimensional) possessing the probability density $f(x)$. Consider statistical estimators $f_n(x)$ of the density $f(x)$ of the form

$$f_n(x) = n^{-1} \sum_{i=1}^{n} \delta_m(x, X_i), \tag{1}$$

*This ADDENDUM was written by E. A. Nadaraya jointly with R. M. Absava and was published in abbreviated form in [104] and [105].

where $\delta_m(x, y)$ is a Borel measurable function and $m = m(n)$ is a sequence of positive numbers approaching infinity but satisfying the condition $m(n) = o(n)$ as $n \to \infty$.

The following notation will be used (here $r(x)$ is a weight function):

$$U_n = n \int (f_n(x) - Ef_n(x))^2 r(x)\, dx, \quad \Delta_n = EU_n,$$

$$\alpha_m(x, y) = \delta_m(x, y) - E\delta_m(x, X_j),$$

$$\delta_{0m}(u, v) = \int \delta_m(x, u)\delta_m(x, v) r(x)\, dx,$$

$$\sigma_n^2 = 2 \int\int (E\alpha_m(u_1, X_1)\alpha_m(u_2, X_1))^2 r(u_1) r(u_2)\, du_1\, du_2,$$

$$\lambda_n = E[\int\int \alpha_m(u_1, X_1)\alpha_m(u_2, X_2) E(\alpha_m(u_1, X_1)\alpha_m(u_2, X_1)) \times$$

$$\times\, r(u_1) r(u_2)\, du_1\, du_2]^2,$$

$$\eta_{ij}^{(n)} = \sqrt{\frac{n}{n-1}}\ \frac{2}{n\sigma_n} \int \alpha_m(x, X_i)\alpha_m(x, X_j) r(x)\, dx,$$

$$\xi_k^{(n)} = \sum_{i=1}^{k-1} \eta_{ik}^{(n)}, \quad k = \overline{2, n}, \quad \xi_1^{(n)} = 0, \quad \xi_k^{(n)} = 0,$$

$$k > n,$$

$$Y_k^{(n)} = \sum_{i=1}^{k} \xi_i^{(n)}, \quad F_k^{(n)} = \sigma(\omega: X_1, \ldots, X_k),$$

where $F_k^{(n)}$ is a σ-algebra generated by the random variables $X_1, \ldots, X_k$, $F_0^{(n)} = (\phi, \Omega)$.

1. Limiting Distribution of U_n

The basic propositions are stated as follows:

1° $\quad \sup_v \int \delta_m^2(u, v) r(u)\, du \le C_1 m^{\alpha}, \quad \alpha > 0;$

2° $\quad \sup_v \int \delta_{0m}^s(u, v) f(u)\, du \le C_2 m^{(s-1)\alpha}, \quad s = 2;\ 4;$

3° $\quad \sigma_n \to \infty$ and $\lambda_n = o(\sigma_n^2);$

4° $\quad n^{-1}\sigma_n^{-4} m^{3\alpha} \to 0$ for $n \to \infty$.

LEMMA 1. The stochastic sequence $(Y_j^{(n)}, F_j^{(n)})$, $j \geq 1$, is a martingale and the sequence $(\xi_j^{(n)}, F_j^{(n)})$, $j \geq 1$, is a difference-martingale.

The proof follows from the representation

$$E(Y_{j+1}^{(n)}/F_j^{(n)}) = E\left(\sum_{i=1}^{j+1} \xi_i^{(n)}/F_j^{(n)}\right) = E\left(\sum_{i=1}^{j} \xi_i^{(n)}/F_j^{(n)}\right) +$$

$$+ E(\xi_{j+1}^{(n)}/F_j^{(n)}) = Y_j^{(n)} \quad \text{a.s.}$$

since $E(\xi_{j+1}^{(n)}|F_j^{(n)}) = 0$ and in view of condition 0 for all $j \geq 1$, $E|Y_j^{(n)}| < \infty$.

Next, since

$$\xi_{j+1}^{(n)} = Y_{j+1}^{(n)} - Y_j^{(n)} \quad \text{and} \quad E(\xi_{j+1}^{(n)}/F_j^{(n)}) = 0$$

a.s., it follows that $(\xi_j^{(n)}, F_j^{(n)})$ $j \geq 1$, is a difference-martingale.

THEOREM 1. Let the conditions 1°-4° be fulfilled. Then as $n \to \infty$

$$\sigma_n^{-1}(U_n - \Delta_n) \overset{d}{\to} N(0, 1),$$

where d denotes the convergence in distribution and $N(a, \sigma)$ denotes a normal random variable with expectation a and variance σ^2.

Proof. We have

$$\frac{U_n - \Delta_n}{\sigma_n} = \frac{1}{n\sigma_n} \sum_{i \neq j} \int \alpha_m(x, X_i)\alpha_m(x, X_j)r(x)\,dx + \frac{1}{n\sigma_n} \sum_{j=1}^{n} \times$$

$$\times \left(\int \alpha_m^2(x, X_j)r(x)\,dx - \Delta_n\right) = \sqrt{\frac{n-1}{n}}\, H_n^{(1)} + H_n^{(2)}, \tag{2}$$

where

$$H_n^{(1)} = \sum_{j=1}^{n} \xi_j^{(n)},$$

$$H_n^{(2)} = \frac{1}{n\sigma_n} \sum_{j=1}^{n} \left(\int \alpha_m^2(x, X_j)r(x)\,dx - E\int \alpha_m^2(x, X_j)r(x)\,dx\right).$$

First we shall verify that $H_n^{(2)}$ converges to zero in probability. Note that conditions 1° and 2° imply the inequalities

$$\sup_v \int \alpha_m^2(u, v)r(u)\,du \leq C_3 m^{\alpha}, \tag{3}$$

$$\sigma_n^2 \leq \int\int \delta_{om}^2(u, v) f(u) f(v)\, du\, dv \leq C_4 m^{\alpha}. \tag{4}$$

From (3) we obtain

$$DH_n^{(2)} = \frac{1}{n\sigma_n^2} D \int \alpha_m^2(x, X_1) r(x)\, dx \leq \frac{1}{n\sigma_n^2} E\left(\int \alpha_m^2(x, X_1) r(x)\, dx\right)^2 = = O\left(\frac{m^{2\alpha}}{n\sigma_n^2}\right)$$

(where D denotes the variance operator). Whence in view of condition 4° and relation (4) $H_n^{(2)} \overset{p}{\to} 0$ as $n \to \infty$ (here and below the letter p above an arrow indicates convergence in probability).

We shall now show that $H_n^{(1)} \overset{d}{\to} N(0, 1)$. For this purpose we verify the validity of Corollaries 2 and 6 of Theorem 2 presented in [103]. It is required to verify the conditions for asymptotic normality of a square integrable martingale-difference stipulated in these assertions. Recall that in view of Lemma 1 our sequence $(\xi_k^{(n)}, F_k^{(n)})$, $k \geq 1$, is such a martingale. Observe that $\sum_{k=1}^{n} E(\xi_k^{(n)})^2 = 1$ since as it is easy to see

$$E(\xi_k^{(n)})^2 = 2(k-1)[n(n-1)]^{-1}.$$

Asymptotic normality will be valid provided for $n \to \infty$

$$\sum_{k=1}^{n} E[(\xi_k^{(n)})^2 I(|\xi_k^{(n)}| \geq \varepsilon)/F_{k-1}^{(n)}] \overset{p}{\to} 0, \tag{*}$$

and

$$\sum_{k=1}^{n} (\xi_k^{(n)})^2 \overset{p}{\to} 1. \tag{**}$$

It was proved in [103] that under (**) and the condition $\sup_{1\leq k\leq n} |\xi_k^{(n)}| \overset{p}{\to} 0$ the condition (*) is satisfied. Since for $\varepsilon > 0$

$$P\left\{\sup_{1\leq k\leq n} |\xi_k^{(n)}| \geq \varepsilon\right\} \leq \varepsilon^{-4} \sum_{k=1}^{n} E(\xi_k^{(n)})^4$$

it follows from relation (9) presented below that to prove the convergence $H_n^{(1)} \overset{d}{\to} N(0, 1)$ it remains to verify that (**) is valid.

Below to simplify the notation we shall write in place of $\alpha_m(x, y)$, $\eta_{ij}^{(n)}$, $\xi_j^{(n)}$ and $\delta_{0m}(x, y)$ simply $\alpha(x, y)$, η_{ij}, ξ_j and $\delta_0(x, y)$.

We shall now prove that $\sum_{k=1}^{n} \xi_k^2 \xrightarrow{P} 1$. For this purpose it is sufficient to verify that

$$E\left(\sum_{k=1}^{n} \xi_k^2 - 1\right)^2 \to 0 \quad \text{as} \quad n \to \infty,$$

i.e. in view of $\sum_{k=1}^{n} E\xi_k^2 = 1$, it is sufficient to show that

$$E\left(\sum_{k=1}^{n} \xi_k^2\right)^2 = \sum_{k=1}^{n} E\xi_k^4 - 2 \sum_{1 \le k_1 < k_2 \le n} E\xi_{k_1}^2 \xi_{k_2}^2 \to 1 \tag{5}$$

as $n \to \infty$.

We shall prove (5). Recalling the definitions of η_{ik} and ξ_k we have

$$\sum_{k=1}^{n} E\xi_k^4 = \frac{16}{n^4 \sigma_n^4} (M_n^{(1)} + M_n^{(2)}), \tag{6}$$

where

$$M_n^{(1)} = \sum_{k=1}^{n} (k-1) \int\int\int\int E \prod_{j=1}^{4} \alpha(x_j, X_1) E \prod_{j=1}^{4} \times$$

$$\times \alpha(x_j, X_k) r(x_j)\, dx_j,$$

$$M_n^{(2)} = 3 \sum_{k=1}^{n} (k-1)(k-2) \int\int\int\int E\alpha(x_1, X_1)\alpha(x_2, X_1) \times$$

$$\times E\alpha(x_3, X_2)\alpha(x_4, X_2) E \prod_{j=1}^{4} \alpha(x_j, X_k) r(x_j)\, dx_j.$$

We now bound $M_n^{(1)}$ and $M_{(n)}^{(2)}$. In view of condition 2° we have

$$M_n^{(1)} = \sum_{k=1}^{n} (k-1) \int\int\int\int (E \prod_{j=1}^{4} \alpha(x_j, X_1))^2 r(x_j)\, dx_j =$$

$$= \sum_{k=1}^{n} (k-1) \int\int (\int \alpha(x_1, u)\alpha(x_1, v) r(x_1)\, dx_1)^4 f(u) \times$$

$$\times f(v)\, du\, dv \le C_5 n^2 \int\int [\delta_0^4(u, v) + (\int f(t)\delta_0(u, t) \times$$

$$\times dt)^4 + (\int f(t)\delta_0(v, t)\, dt)^4 + (\int\int \delta_0(x, y) f(x) \times$$

$$\times f(y)\, dx\, dy)^4] f(u) f(v)\, du\, dv \le C_6 n^2 m^{3\alpha}, \tag{7}$$

and also

$$M_n^{(2)} = 3 \sum_{k=1}^{n} (k-1)(k-2) \int\int\int (\int \alpha(x_1, u)\alpha(x_1, t) r(x_1) \times$$

$$\times\, dx_1)^2 (\int \alpha(x_2, v)\alpha(x_2, t) r(x_2)\, dx_2)^2 f(u) f(v) f(t) \times$$

$$\times\, du\, dv\, dt \leq 3 \sum_{k=1}^{n} (k-1)(k-2) \int\int\int (\delta_0^2(u, t) +$$

$$+ O(m^{\alpha}))({}_0^2(v, t) + O(m^{\alpha})) f(u) f(v) f(t)\, du\, dv\, dt \leq C_7 n^3 m^{2\alpha}. \tag{8}$$

It follows from (6) and (8) that

$$\sum_{k=1}^{n} E\xi_k^4 \leq C_8 \frac{m^{3\alpha}}{n\sigma_n^4}.$$

Hence in view of condition 4°

$$\sum_{k=1}^{n} E(\xi_k^4) \to 0 \quad \text{as} \quad n \to \infty. \tag{9}$$

Next it follows from the definiton of ξ_i that

$$\xi_{k_1}^2 \xi_{k_2}^2 = \left(\sum_{i=1}^{k_1-1} \eta_{ik_1}^2 \right) \left(\sum_{i=1}^{k_2-1} \eta_{ik_2}^2 \right) + \left(\sum_{i=1}^{k_1-1} \eta_{ik_1}^2 \right) \left(\sum_{i\neq s=1}^{k_2-1} \times \right.$$

$$\left. \times\, \eta_{ik_2}\eta_{sk_2} \right) + \left(\sum_{i=1}^{k_2-1} \eta_{ik_2}^2 \right) \left(\sum_{s\neq t=1}^{k_1-1} \eta_{sk_1}\eta_{tk_1} \right) +$$

$$+ \left(\sum_{s\neq t=1}^{k_1-1} \eta_{sk_1}\eta_{tk_1} \right) \left(\sum_{k\neq r=1}^{k_2-1} \eta_{kk_2}\eta_{rk_2} \right) =$$

$$= B_{k_1k_2}^{(1)} + B_{k_1k_2}^{(2)} + B_{k_1k_2}^{(3)} + B_{k_1k_2}^{(4)}.$$

Hence

$$2 \sum_{1\leq k_1<k_2\leq n} E\xi_{k_1}^2 \xi_{k_2}^2 = \sum_{i=1}^{4} A_n^{(i)},$$

where

$$A_n^{(i)} = 2 \sum_{1\leq k_1<k_2\leq n} EB_{k_1k_2}^{(i)}, \quad i = \overline{1, 4}.$$

First we shall consider the sum $A_n^{(3)}$; utilizing the definitions of η_{ij} and $\alpha(x, y)$ we have

$$E\eta_{ik_2}^2 \eta_{sk_1} \eta_{tk_1} = 0, \quad s \neq t, \quad k_1 < k_2.$$

Whence we obtain

$$A_n^{(3)} = 0. \tag{10}$$

We now bound $A_n^{(2)}$. For this purpose we write $EB_{k_1k_2}^{(2)}$ in the form of a sum

$$EB_{k_1k_2}^{(2)} = \sum_{i=1}^{k_1-1} \sum_{r\neq s=1}^{k_1} E\eta_{ik_1}^2 \eta_{rk_2} \eta_{sk_2} + \sum_{i=1}^{k_1-1} \sum_{r\neq s=k_1+1}^{k_2-1} E\eta_{ik_1}^2 \eta_{rk_2} \eta_{sk_2} \tag{11}$$

The second summand in (11) equals zero since i cannot be equal to either r or s and $r \neq s$, under these restrictions

$$E\eta_{ik_1}^2 \eta_{rk_2} \eta_{sk_2} = 0.$$

In the first summand $E\eta_{ik_1}^2 \eta_{rk_2} \eta_{sk_2} = 0$ except when $s = k_1$ or $r = k_1$. Thus

$$EB_{k_1k_2}^{(2)} = 2 \sum_{i=1}^{k_1-1} E(\eta_{ik_1}^2 \eta_{ik_2} \eta_{k_1k_2}) \tag{12}$$

In view of (3) we arrive at the bound

$$|EB_{k_1k_2}^{(2)}| \leq \frac{32(k_1 - 1)}{n^4\sigma_n^4} E\left(\int \alpha(u, X_1)\alpha(u, X_2)r(u)\, du\right)^2 \left| \int \alpha \times \right.$$

$$\left. \times (v, X_1)\alpha(v, X_3)r(v)\, dv \right| \left| \int \alpha(t, X_2)\alpha(t, X_3)r(t)\, dt\right| \leq$$

$$\leq C_9(k_1 - 1)\sigma_n^2 m^{2\alpha} n^{-4}.$$

Whence

$$|A_n^{(2)}| \leq C_{10} \frac{m^{2\alpha}}{n^4\sigma_n^2} \sum_{k_1<k_2} (k_1 - 1) = O\left(\frac{m^{2\alpha}}{n\sigma_n^2}\right) \tag{13}$$

Consider now $A_n^{(4)}$. We have

$$EB_{k_1k_2}^{(4)} = E \sum_{s\neq t=1}^{k_1-1} \eta_{sk_1}\eta_{tk_1} \sum_{k\neq r=1}^{k_2-1} \eta_{kk_2}\eta_{rk_2} = 4 \sum_{s<t}^{k_1-1} \sum_{k<r}^{k_1-1} \times$$

$$\times E\eta_{sk_1}\eta_{tk_1}\eta_{kk_2}\eta_{rk_2}.$$

In this case the summands of the form $E\eta_{sk_1}\eta_{tk_1}\eta_{sk_2}\eta_{tk_2}$ may be different from zero while all the others are zero. Therefore

$$EB^{(4)}_{k_1k_2} = 4 \sum_{s<t}^{k_1-1} E\eta_{sk_1}\eta_{tk_1}\eta_{sk_2}\eta_{tk_2} . \tag{14}$$

Evidently,

$$E\eta_{sk_1}\eta_{tk_1}\eta_{sk_2}\eta_{tk_2} = E(E\eta_{sk_1}\eta_{tk_1}\eta_{sk_2}\eta_{tk_2} \mid X_t, X_{k_1}, X_{k_2})) =$$

$$= E(\eta_{tk_1}\eta_{tk_2}E(\eta_{sk_1}\eta_{sk_2} \mid X_t, X_{k_1}, X_{k_2})).$$

However,

$$E(\eta_{sk_1}\eta_{sk_2} \mid X_t, X_{k_1}, X_{k_2}) = \frac{n}{n-1}\frac{4}{n^2\sigma_n^2} \int\int \alpha(x, X_{k_1}) \times$$

$$\times \alpha(y, X_{k_2})E(\alpha(x, X_s)\alpha(y, X_s) \mid X_t, X_{k_1}, X_{k_2})r(x)r(y)\,dx\,dy =$$

$$= \frac{4n}{n-1}\frac{1}{n^2\sigma_n^2} \int\int \alpha(x, X_{k_1})\alpha(y, X_{k_2})E\alpha(x, X_1)\alpha(y, X_1) \times$$

$$\times r(x)r(y)\,dx\,dy.$$

Hence,

$$E\eta_{sk_1}\eta_{tk_1}\eta_{sk_2}\eta_{tk_2} = \frac{4n}{n-1}\frac{1}{n^2\sigma_n^2} E\eta_{tk_1}\eta_{tk_2} \int\int \alpha(x, X_{k_1})\alpha(y, X_{k_2}) \times$$

$$\times E\alpha(x, X_1)\alpha(y, X_1)r(x)r(y)\,dx\,dy = \left(\frac{n}{n-1}\right)^2 \frac{16}{n^4\sigma_n^4} \times$$

$$\times E \int\int \alpha(u, X_{k_1})\alpha(v, X_{k_2})E\alpha(u, X_t)\alpha(v, X_t)r(u)r(v)\,du\,dv \times$$

$$\times \int\int \alpha(x, X_{k_1})\alpha(y, X_{k_2})E\alpha(x, X_1)\alpha(y, X_1)r(x)r(y)\,dx\,dy =$$

$$= \left(\frac{n}{n-1}\right)^2 \frac{16}{n^4\sigma_n^4} E[\int\int \alpha(u, X_1)\alpha(v, X_2)E\alpha(u, X_1) \times$$

$$\times \alpha(v, X_1)r(u)r(v)\,du\,dv]^2 = \left(\frac{4n}{n-1}\right)^2 \frac{1}{n^4\sigma_n^4}\lambda_n.$$

From here taking condition 2° into account we have

$$|E\eta_{sk_1}\eta_{tk_1}\eta_{sk_2}\eta_{tk_2}| \le C_{11}\frac{1}{n^4\sigma_n^2}. \tag{15}$$

We thus obtain from (14) and (15) that

$$|A_n^{(4)}| \le C_{12}\frac{1}{n^4\sigma_n^2}\sum_{k_1<k_2}(k_1-1)(k_1-2) \le C_{13}\frac{1}{\sigma_n^2},$$

since

$$\sum_{k_1<k_2}(k_1-1)(k_1-2) = O(n^4).$$

Since by the condition $\sigma_n \to \infty$ as $n \to \infty$ we obtain from here that

$$A_n^{(4)} \to 0 \quad \text{as} \quad n \to \infty. \tag{16}$$

Finally we shall show that $A_n^{(1)} \to 1$ as $n \to \infty$.

We have

$$EB_{k_1k_2}^{(1)} = \sum_{k=1}^{k_1-1} E\eta_{kk_1}^2\eta_{kk_2}^2 + \sum_{k\ne s} E\eta_{kk_1}^2\eta_{sk_2}^2 + \sum_{k=1}^{k_1-1} E\eta_{kk_1}^2\eta_{kk_2}^2 +$$

$$+ \sum_{k=1}^{k_1-1}\sum_{s=k_1+1}^{k_2-1} E\eta_{kk_1}^2\eta_{sk_2}^2 = 2(k_1-1)E\eta_{12}^2\eta_{13}^2 + (k_1-1)\times$$

$$\times (k_1-2)(E\eta_{12}^2) + (k_1-1)(k_2-k_1-1)(E\eta_{12}^2)^2 =$$

$$= 2(k_1-1)E\eta_{12}^2\eta_{13}^2 + (k_1-1)(k_2-3)(E\eta_{12}^2)^2.$$

Hence

$$A_n^{(1)} = 4E\eta_{12}^2\eta_{13}^2\sum_{k_1<k_2}(k_1-1) + 2(E\eta_{12}^2)^2\sum_{k_1<k_2}(k_1-1)(k_2-3). \tag{17}$$

Using the relations

$$\sum_{k_1<k_2}(k_1-1)(k_2-3) = \frac{n(n-1)(n-2)(n-3)}{8},$$

$$\sum_{k_1<k_2}(k_1-1) = \frac{n(n-1)(n-2)}{6},$$

$$E\eta_{12}^2\eta_{13}^2 = \left(\frac{4n}{n-1}\right)^2\frac{1}{n^4\sigma_n^4}E(\int \alpha(u, X_1)\alpha(u, X_2)r(u)\,du)^2 \times$$

$$\times \left(\int \alpha(v, X_1)\alpha(v, X_3)r(v)\, dv\right)^2 \le C_{14} \frac{m^{2\alpha}}{n^4\sigma_n^4},$$

$$(E\eta_{12}^2)^2 = \left(\frac{2n}{n-1}\right)^2 \frac{1}{n^4\sigma_n^4} \left(2 \int\int (E\alpha(u, X_1)\alpha(v, X_1))^2 r(u)r(v) \times\right.$$

$$\left. \times\, du\, dv\right)^2 = \left(\frac{2n}{n-1}\right)^2 \frac{\sigma_n^4}{n^4\sigma_n^4} = \left(\frac{2n}{n-1}\right)^2 \frac{1}{n^4}$$

we obtain from (17) that

$$A_n^{(1)} = O\left(\frac{m^{2\alpha}}{n\sigma_n^4}\right) + 1 + O(1).$$

Hence

$$A_n^{(1)} \to 1 \quad \text{as} \quad n \to \infty. \tag{18}$$

Combining the relations (9), (10), (13), (16) and (18) we finally arrive at

$$E\left(\sum_{k=1}^{n} \xi_k^2 - 1\right)^2 \to 0 \quad \text{as} \quad n \to \infty.$$

The theorem is thus proved.

2. Kernel Density Estimators/Rosenblatt-Parzen Estimators

Let $X_i = (X_i^{(1)}, \ldots, X_i^{(p)})$, $i = \overline{1, n}$ be a sequence of independent identically distributed random variables with the values in the Euclidean p-dimensional space R_p possessing the density $f(x)$, $x = (x_1, \ldots, x_p)$.

In this section we shall assume that

$$\delta_m(x, y) = m^p K(m(x - y)), \quad x, y \in R_p.$$

Then

$$f_n(x) = \frac{m^p}{n} \sum_{j=1}^{n} K(m(x - X_j)).$$

A functions $K(x)$, $x \in R_p$, will be called a function of the class H_s, $(s \ge 2)$, provided it satisfies the following conditions:

$$\int K(x)\, dx = 1, \qquad \sup_{x \in R_p} |K(x)| < \infty; \int x_1^{i_1} \ldots x_p^{i_p} K(x)\, dx = 0$$

$$\text{if } \sum_{j=1}^{p} i_j < s; \quad \int |x_1^{i_1} \dots x_p^{i_p} K(x)|\, dx < \infty \quad \text{if } \sum_{j=1}^{p} i_j = s.$$

Denote by W_s the set of all bounded functions possessing bounded partial derivatives up to the s-th, $s \geq 2$, order inclusively.

Let $r(x) = r_1(x_1) \cdot \dots \cdot r_p(x_p)$, $r_j(x_j) \geq 0$, x_j R_1, be piecewise-continuous bounded and integrable functions.

Conditions 1° and 2° of Theorem 1 are satisfied in this case for $\alpha = p$. Indeed we have

1° $$\sup_v \int \delta_m^2(u, v) r(u)\, du = m^{2p} \sup_v \int K^2(m(u - v)) r(u)\, du \leq$$
$$\leq C_{15}\, m^p \int K^2(u)\, du.$$

2° $$\sup_v \int \delta_0^s(u, v) f(u)\, du \leq C_{16} m^{sp} \sup_v \int \bar{K}_0^s(m(u - v))\, du \leq$$
$$\leq C_{16} m^{(s-1)p} \int \bar{K}_0^s(u)\, du, \quad s = 2; 4,$$

where

$$\bar{K}_0(u) = \int |K(t)|\, |K(u - t)|\, dt.$$

We shall now verify the validity of condition 3°.

Since $\alpha(u, v) = m^p[K(m(u - v)) - EK(m(u - X_1))]$,

we have

$$\sigma_n^2 = 2 \int\int (E\alpha(u_1, X_1)\alpha(u_2, X_1))^2 r(u_1) r(u_2)\, du_1\, du_2 =$$
$$= 2m^{4p} \int\int [m^{-p} \int K(t) K(m(u_2 - u_1) - t) f(u_1 - \tfrac{t}{m})\, dt -$$
$$- m^{-2p} \int K(t) f(u_1 - \tfrac{t}{m})\, dt \int K(t) f(u_2 - \tfrac{t}{m})\, dt]^2 \times$$
$$\times r(u_1) r(u_2)\, du_1\, du_2.$$

Expanding $f(u_1 - \frac{t}{m})$ in accordance with Taylor's formula we obtain from here

$$\sigma_n^2 = 2m^{4p} \int\int [m^{-2p} K_0^2(m(u_2 - u_1)) f^2(u_1) + O(m^{-2p-2} \Big(\sum_{j=1}^{p} \int t_j \times$$
$$\times K(t) K(m(u_2 - u_1) - t))\, dt)^2) + O(m^{2p-1} K_0(m(u_2 - u_1)) \times$$
$$\times \sum_{j=1}^{p} \int t_j K(t) K(m(u_2 - u_1) - t)\, dt + O(m^{-3p} K_0(m(u_2 - u_1)))$$

$$+ O(m^{-3p-1} \sum_{j=1}^{p} \int t_j K(t)K(m(u_2 - u_1)) - t)\, dt + O(m^{-4p})] \times$$

$$\times\, r(u_1)r(u_2)\, du_1\, du_2, \tag{19}$$

where

$$K_0(x) = \int K(t)K(x - t)\, dt, \quad x \in R_p.$$

In view of Fubini's theorem we obtain from (19) that

$$\sigma_n^2 = 2m^{2p} \int\int K_0^2(m(u_2 - u_1))f^2(u_1)r(u_2)\, du_1\, du_2 + O(1). \tag{20}$$

It is known [33] that $m^p \int K_0^2(m(u_2 - u_1))f^2(u_1)r(u_1)\, du_1 \to f^2(u_2)r(u_2) \times \int K_0^2(t)\, dt$ as $n \to \infty$. Therefore, utilizing Lebesgue's théorem on limiting transition under the integral sign it follows from (20) that

$$m^{-p}\sigma_n^2 \to \sigma^2 = 2 \int f^2(u)r^2(u)\, du \int K_0^2(v)\, dv \tag{21}$$

as $n \to \infty$. Thus $\sigma_n^2 \to \infty$ as $n \to \infty$, i.e. the first part of condition 3° is fulfilled. It remains to verify the second part of this condition, i.e. to show that $\lambda_n = O(\sigma_n^2)$.

We have

$$\lambda_n = E[\int\int \alpha(u_1, X_1)\alpha(u_2, X_2)E\alpha(u_1, X_1)\alpha(u_2, X_1)r(u_1)r(u_2) \times$$

$$\times\, du_1\, du_2]^2 = m^{8p} \int\int\int\int \operatorname{cov}(K(m(u_1 - X_1)),$$

$$K(m(v_1 - X_1))\operatorname{cov}(K(m(u_2 - X_2)),$$

$$K(m(u_2 - X_1))\operatorname{cov}(K(m(v_1 - X_1)),$$

$$K(m(v_2 - X_1))r(u_1)r(u_2)r(v_1)r(v_2)\, du_1\, du_2\, dv_1\, dv_2 =$$

$$= m^{8p} \int\int\int\int [\int K(m(u_1 - t_1))K(m(v_1 - t_1))f(t_1)\, dt_1 -$$

$$- EK(m(u_1 - X_1))EK(m(v_1 - X_1))] \cdot [\int K(m(u_2 - t_2))K(m \times$$

$$\times (v_2 - t_2))f(t_2)\, dt_2 - EK(m(u_2 - X_1))EK(m(v_2 - X_1))] \times$$

$$\times [\int K(m(u_1 - t_3))K(m(u_2 - t_3))f(t_3)\, dt_3 - EK(m(u_1 - X_1)) \times$$

$$\times \quad EK(m(u_2 - X_1))][\int K(m(v_1 - t_4))K(m(v_2 - t_4))f(t_4)\, dt_4 -$$

$$- EK(m(v_1 - X_1))EK(m(v_2 - X_1))] r(u_1)r(u_2)r(v_1)r(v_2) \times$$
$$\times du_1\, du_2\, dv_1\, dv_2. \tag{22}$$

Since

$$\int |K(m(u - t)| K(m(v - t))|f(t)\, dt \le Mm^{-p}\bar{K}_0(m(v - u))$$

and

$$|EK(m(u - X_1))| \le Mm^{-p},$$

where $M = \sup_{x \in R_p} f(x)$ we obtain from (22) that

$$\lambda_n \le M^4 m^{4p} \int\int\int\int [\bar{K}_0(m(v_1 - u_1)) + m^{-p}][\bar{K}_0(m(v_2 - u_2)) + m^{-p}] \times$$
$$\times [\bar{K}_0(m(u_2 - u_1)) + m^{-p}][\bar{K}_0(m(v_2 - v_1)) + m^{-p}]r(u_1)r(u_2) \times$$
$$\times\ r(v_1)r(v_2)\, du_1\, du_2\, dv_1\, dv_2. \tag{23}$$

The integral in (23) can be represented as a sum of 16 integrals - they coincide with one of the five possible integrals of the following type:

$$L_1 = \int\int\int\int \bar{K}_0(m(v_1 - u_1))\bar{K}_0(m(v_2 - u_2))\bar{K}_0(m(u_2 - u_1))\bar{K}_0 \times$$
$$\times (m(v_2 - v_1))r(u_1)r(u_2)r(v_1)r(v_2)\, du_1\, du_2\, dv_1\, dv_2,$$

$$L_2 = m^{-p} \int\int\int\int \bar{K}_0(m(v_1 - u_1))\bar{K}_0(m(u_2 - u_1))\bar{K}_0(m(v_2 - v_1)) \times$$
$$\times r(u_1)r(u_2)r(v_1)r(v_2)\, du_1\, du_2\, dv_1\, dv_2,$$

$$L_3 = m^{-2p} \int\int\int\int \bar{K}_0(m(v_1 - u_1))\bar{K}_0(m(v_2 - v_1))r(u_1)r(u_2)r(v_1) \times$$
$$\times\ r(v_2)\, du_1\, du_2\, dv_1\, dv_2,$$

$$L_4 = m^{-3p} \int\int\int\int \bar{K}_0(m(v_2 - v_1))r(u_1)r(u_2)r(v_1)r(v_2)\, du_1\, du_2 \times$$
$$\times\ dv_1\, dv_2,$$

$$L_5 = m^{-4p} \int\int\int\int r(u_1)r(u_2)r(v_1)r(v_2)\, du_1\, du_2\, dv_1\, dv_2 = O(m^{-4p}).$$

Simple calculations show that

$$L_1 = O(m^{-3p}), \quad L_2 = O(m^{-3p}), \quad L_3 = O(m^{-4p}), \quad L_4 = O(m^{-3p}),$$

and $$L_5 = O(m^{-4p}). \tag{24}$$

Thus from (23) and (24) we arrive at

$$\lambda_n \leq c_{17} m^p;$$

this together with (21) implies that

$$\lambda_n = O(\sigma_n^2). \tag{25}$$

Condition 3° is therefore fully verified. Next taking (21) into account condition 4° becomes

$$n^{-1} m^p \to 0 \quad \text{as} \quad n \to \infty. \tag{26}$$

Combining the relations obtained in (21), (25) and (26) we arrive at the following theorem.

THEOREM 2. Let $K \in H_s$ and $f \in W_s$. If $n^{-1}m^p \to 0$ as $n \to \infty$ then

$$m^{p/2}\sigma^{-1}(\bar{U}_n - \bar{\Delta}_n) \xrightarrow{d} N(0, 1),$$

where

$$\bar{U}_n = \frac{n}{m^p} \int (f_n(x) - Ef_n(x))^2 r(x)\, dx, \qquad \bar{\Delta}_n = E\bar{U}_n.$$

More natural for applications is of course the situation when $Ef_n(x)$ in $\bar{U}_n$ is replaced by $f(x)$.

Denote

$$T_n = nm^{-p} \int (f_n(x) - f(x))^2 r(x)\, dx.$$

The following lemma is valid.

LEMMA 2. Let $K \in H_s$ and $f \in W_s$. If $nm^{-p-2s} \to 0$ as $n \to \infty$ then

$$m^{p/2}(T_n - \bar{U}_n - \Theta_n^{(1)}) = o_p(1),$$

where

$$\Theta_n^{(1)} = nm^{-p} \int (Ef_n(x) - f(x))^2 r(x)\, dx.$$

Proof. We have

$$T_n - \bar{U}_n - \Theta_n^{(1)} = R_n = 2nm^{-p} \int (Ef_n(x) - f(x))(f_n(x) - Ef_n(x)) \times$$
$$\times\, r(x)\, dx.$$

Furthermore since $f \in W_s$ and $K \in H_s$ we can write

$$Ef_n(x) - f(x) = m^{-s} \sum_{|\ell|=s} \int \frac{t^\ell}{\ell!} K(t) f^{(\ell)}(x - \Theta \frac{t}{m})\, dt = O(m^{-s}),$$

where

$$|\ell| = \sum_{j=1}^{p} \ell_j, \quad \ell! = \ell_1! \ldots \ell_p!, \quad t^\ell = t_1^{\ell_1} \ldots t_p^{\ell_p},$$

$$f^{(\ell)}(x) = \frac{\partial^{|\ell|} f(x)}{\partial x_1^{\ell_1} \ldots \partial x_p^{\ell_p}}. \tag{27}$$

The bound $O(\cdot)$ is uniform with respect to $x \in R_p$. We now estimate $E|R_n|$. Since

$$\mathrm{cov}(f_n(x),\ f_n(y)) = n^{-1}m^{2p}\Big\{\int K(m(x-u))K(m(y-u))f(u)\,du - $$
$$- \int K(m(x-u))f(u)\,du \int K(m(y-u))f(u)\,du\Big\},$$

we obtain

$$E|R_n| \le 2nm^{-p}\Big\{n^{-1}m^{2p}E[\int K(m(x_1 - X_1))(Ef_n(x_1) - f(x_1))r(x_1) \times$$
$$\times dx_1 - E\int K(m(x_1 - X_1))(Ef_n(x_1) - f(x_1))r(x_1)\,dx_1{}^2\Big\}^{1/2}$$
$$\le 2\sqrt{n}\Big\{\int f(u)\,du(\int K(m(x-u))(Ef_n(x) - f(x)) \times$$
$$\times r(x)\,dx)^2\Big\}^{1/2} \le C_{18}\sqrt{n}\, m^{-(p/2)-s}.$$

The lemma is thus proved.

LEMMA 3. $\bar{\Delta}_n = E\bar{U}_n = \Theta_n^{(2)} + O(m^{-p})$,

where

$$\Theta_n^{(2)} = m^p \int\int K^2(m(x-u))f(u)r(x)\,du\,dx.$$

THEOREM 3. Let $K \in H_s$ and $f \in W_s$. Then

a) If $n^{-1}m^p \to 0$ and $nm^{-p-2s} \to 0$ as $n \to \infty$, then

$$m^{p/2}\sigma^{-1}(T_n - \Theta_n^{(1)} - \Theta_n^{(2)}) \xrightarrow{d} N(0,\ 1);$$

b) If $n^{-1}m^p \to 0$ and $nm^{-(p/2)-2s} \to 0$, then

$$m^{p/2}\sigma^{-1}(T_n - \Theta_n^{(2)}) \xrightarrow{d} N(0,\ 1);$$

c) To the conditions imposed on $K(x)$, $x \in R_p$, we add the condition

$$\int x_j K^2(x)\,dx = 0, \quad j = \overline{1,\ p}.$$

If $n^{-1}m^p \to 0$, $nm^{-(p/2)-2s} \to 0$ as $n \to \infty$ and $p \leq 3$, then

$$m^{p/2}\sigma^{-1}(T_n - \int f(x)r(x)\,dx \int K^2(u)\,du) \xrightarrow{d} N(0, 1).$$

<u>Proof</u>. Assertion a) follows from Theorem 2 and Lemmas 2 and 3 in an obvious manner. Assertion b) follows directly from (26) since

$$m^{p/2}\Theta_n^{(1)} = nm^{-p/2} \int (Ef_n(x) - f(x))^2 r(x)\,dx = O(nm^{-(p/2)-2s}).$$

Finally assertion c) follows from the previous assertion and the following representation of $\Theta_n^{(2)}$:

$$\Theta_n^{(2)} = \int f(x)r(x)\,dx \int K^2(u)\,du + O(m^{-2}).$$

Set $m = n^\alpha$, $\alpha > 0$. Conditions concerning n and m stipulated in a) are satisfied if $(p + 2s)^{-1} < \alpha < p^{-1}$; conditions stipulated in b) are valid if $2(p + 4s)^{-1} < \alpha < p^{-1}$ and $p < 4s$, and the conditions of assertion c) are compatible if $2(p + 4s)^{-1} < \alpha < p^{-1}$ and $p \leq 3$.

As it was mentioned above, in papers [100] and [101] a martingale approach to investigation of the limiting distribution of the quadratic deviation of Rosenblatt-Parzen's estimators was introduced. In [100] the one-dimensional version of Theorem 3 was considered (and the possibility of multi-dimensional generalization of this theorem indicated); in [101] assertion a) of Theorem 3 in the case $s = 2$ was presented.

Next we shall consider an application of Theorem 3 in the two-dimensional case for testing statistical hypotheses of fit.

Let $X_i = (X_i^{(1)}, X_1^{(2)})$, $i = 1, 2, \ldots$ be a sequence of independent identically distributed random variables possessing the density $f(x)$, $x = (x_1, x_2)$.

Assertion c) of Theorem 3 allows us to construct a test at the asymptotic level α for testing the hypothesis H_0 that $f(x) = f_0(x)$. Here we calculate T_n and reject H_0 if

$$T_n \geq d_n(\alpha) = \int f_0(x)r(x)\,dx \int K^2(u)\,du + m^{-1}\lambda_\alpha\sigma,$$

where λ_α is the α-th quantile of the standard normal distribution.

Assume now that the tested hypothesis H_0 is wrong and that actually the hypothesis

$$H^{(n)}: f^{(n)}(x) = f_0(x) + \gamma_n\phi(x), \quad x \in R_2, \quad \int \phi(x)\,dx = 0,$$

$$\gamma_n \downarrow 0.$$

is valid.

THEOREM 4. Let $K(x)$ and $f^{(n)}(x)$ satisfy the conditions of assertion c) of Theorem 3. Let

$$T_n = nm^{-2} \int (f_n(x) - f_0(x))^2 r(x)\, dx.$$

If $m = n^{\delta}$ and $\gamma_n = m^{1/2} n^{-1/2}$, $(1 + 2s)^{-1} < \delta < 2^{-1}$, then

$$m(T_n - \int f_0(x) r(x)\, dx \int K^2(u)\, du) \xrightarrow{d} N(a, \sigma),$$

$$a = \int \phi^2(u) r(u)\, du, \quad \sigma^2 = 2 \int f_0^2(u) r^2(u)\, du \int K_0^2(v)\, dv.$$

COROLLARY. A local behavior of the power $P_{H^{(n)}}(T_n \geq d_n(\alpha))$ is as follows: as $n \to \infty$

$$P_{H^{(n)}}(T_n \geq d_n(\alpha)) \to 1 - \Phi(\lambda_\alpha - \frac{1}{\sigma} \int \phi^2(u) r(u)\, du).$$

It should be noted that for positive weight functions $r(x)$, $x \in R_2$, the test for testing the hypothesis H_0 versus an alternative of the form $H^{(n)}$ is an asymptotically strictly unbiased one since the mathematical expectation of $\int \phi^2(x) r(x)\, dx > 0$ and equals zero if and only if $\phi(x) = 0$, x R_2.

We now consider the 'singular' alternatives introduced in Chapter 3, Section 3:

$$H^{(n)}: f^{(n)}(x_1, x_2) = f_0(x_1, x_2) + \alpha_n \phi\left(\frac{x_1 - \ell_1}{\gamma_n}, \frac{x_2 - \ell_2}{\gamma_n}\right),$$

where $\alpha_n \downarrow 0$, $\gamma_n \downarrow 0$, the function $\phi(u_1, u_2)$ is bounded and possesses bounded partial derivatives up to the second order, and moreover

$$\int \phi(u_1, u_2)\, du_1\, du_2 = 0,$$

$\ell = (\ell_1, \ell_2)$ is a fixed point of continuity of $r(x)$ such that $r(\ell) \neq 0$.

THEOREM 5. Let $K(x)$ and $f_0(x)$, $x \in R_2$, satisfy the conditions of Theorem 4 for $s = 2$. Let $m = n^{\gamma}$, $1/5 < \gamma < 1/2\alpha$, $\alpha_n = m^{\alpha} n^{-1/2}$, $\gamma = m^{-\beta}$, $2\beta > \alpha$, $\alpha - \beta = 1/2$, $1 < \alpha < 7/6$, $1/2 < \beta < 2/3$. Then

$$m(T_n - \int f_0(x) r(x)\, dx \int K^2(u)\, du) \xrightarrow{d} N(a, \sigma),$$

$$a = r(\ell) \int \phi^2(u)\, du, \quad \sigma^2 = 2 \int f_0^2(x) r^2(x)\, dx \int K_0^2(v)\, dv.$$

The proof of Theorem 5 is completely analogous to the proof of Theorem 1 in Chapter 3, Section 3.

The conditions on α and β are satisfied when for example, one sets

$$\alpha = \frac{11}{10}, \quad \beta = \frac{3}{5}; \quad \alpha = \frac{8}{7}, \quad \beta = \frac{9}{14}; \quad \alpha = \frac{9}{8}, \quad \beta = \frac{5}{8};$$

$$\alpha = \frac{10}{9}, \quad \beta = \frac{11}{18} \quad \text{and so on.}$$

Remark. Integrating $f^{(n)}(x, y)$ we verify that the alternatives differ from the null hypothesis by a quantity of order $\alpha_n \gamma_n^2 = o(n^{-1/2})$.

This means that the tests based on the deviation between two-dimensional empirical and hypothetical distribution functions cannot distinguish between the 'singular' and the initial hypothesis. This means that tests based on T_n under 'singular' alternatives are more powerful than the tests based on deviations.

3. Projection Estimators of Probability Density/Chentsov Estimators

Let $X_1, \ldots, X_n$ be independent p-dimensional random vectors whose distribution is given by the probability density $f(x)$, $x \in R_p$. Let $L_2(R_p, r)$ be the space of functions defined on R_p and square integrable with respect to measure μ, $d\mu = r(x)\,dx$. Select in the space an orthonormal basis $\{\phi_j(x)\}$. Next let

$$\delta_m(x, y) = \sum_{j=1}^{m} \phi_j(x)\phi_j(y)r(y).$$

Then from (1) we obtain Chentsov's estimator:

$$f_n(x) = \sum_{j=1}^{m} \hat{a}_j \phi_j(x),$$

where

$$\hat{a}_j = n^{-1} \sum_{k=1}^{n} \phi_j(X_k)r(X_k).$$

The quantities $\Delta_n = EU_n$ and σ_n^2 will now be expressed explicitly in terms of the basis $\{\phi_j\}$.

We have

$$\Delta_n = En \int (f_n(x) - Ef_n(x))^2 r(x)\,dx = \int E[\delta_m(u, X_1) - E\delta_m(u, X_1)]^2 r(u)\,du = \sum_{j=1}^{m} E[\phi_j(X_1)r(X_1) - E\phi_j(X_1)r(X_1)]^2 =$$

$$= \sum_{j=1}^{m} [\int \phi_j^2(u)r^2(u)f(u)\,du - a_j^2], \tag{28}$$

where

$$a_j = \int \phi_j(x)r(x)f(x)\,dx.$$

$$\sigma_n^2 = 2 \int\int (E\alpha(u_1, X_1)\alpha(u_2, X_1))^2 r(u_1)r(u_2)\,du_1\,du_2 =$$

$$= 2 \int\int [E\delta_m(u_1, X_1) - E\delta_m(u_1, X_1))(\delta_m(u_2, X_1) - E\delta_m \times$$
$$\times (u_2, X_1))]^2 r(u_1)r(u_2)\, du_1\, du_2 = 2 \sum_{i=1}^{m} \sum_{j=1}^{m} [\mathrm{cov}(\phi_i(X_1) \times$$
$$\times r(X_1),\ \phi_j(X_1)r(X_1))]^2 = 2 \sum_{i=1}^{m} \sum_{j=1}^{m} (\int \phi_i(u)\phi_j(u)r^2(u) \times$$
$$\times f(u)\, du - a_i a_j)^2. \tag{29}$$

Now let $M = \sup_{x \in R_p} f(x)r(x) < \infty$. We shall show that

$$\lambda_n = O(\sigma_n^2). \tag{30}$$

Indeed

$$\lambda_n = E[\int\int \alpha(u_1, X_1)\alpha(u_2, X_2)E\alpha(u_1, X_1)\alpha(u_2, X_1)r(u_1)r(u_2) \times$$
$$\times du_1\, du_2]^2 = E[\sum_{i,j} \phi_i^*(X_1)\,\phi_j^*(X_2)a_{ij}]^2, \tag{31}$$

where

$$\phi_i^*(X_1) = \phi_i(X_1)r(X_1) - E\phi_i(X_1)r(X_1),$$

$$a_{ij} = \mathrm{cov}(\phi_i(X_1)r(X_1),\quad \phi_j(X_1)r(X_1)).$$

From (31) we have

$$\lambda_n \leq 4[E(\sum_{i,j} a_{ij}\phi_i(X_1)\phi_j(X_2)r(X_1)r(X_2))^2 + 2E(\sum_{i,j} a_{ij}a_i \times$$
$$\times \phi_j(X_1)r(X_1))^2 + (\sum_{i,j} a_{ij}a_i a_j)^2] = \Sigma_1 + \Sigma_2 + \Sigma_3. \tag{32}$$

We estimate Σ_1. In view of $M = \sup_{x \in R_p} f(x)r(x) < \infty$ and (29) we have

$$\Sigma_1 = 4E(\sum_{i,j} a_{ij}\phi_i(X_1)\phi_j(X_2)r(X_1)r(X_2))^2 = 4\int\int(\sum_{i,j} a_{ij}\phi_i \times$$
$$\times (u)\phi_j(v)r(u)r(v))^2 f(u)f(v)\, du\, dv \leq 4M^2 \int\int (\sum_{i,j} a_{ij}\phi_i \times$$
$$\times (u)\phi_j(v))^2 r(u)r(v)\, du\, dv = 4M^2 \sum_{i,j} a_{ij}^2 = 2M^2\sigma_n^2. \tag{33}$$

It is easy to estimate Σ_2 since

$$\Sigma_2 = 8E(\sum_{i,j} a_{ij}a_i\phi_j(X_1))^2 = 8\int(\sum_{i,j} a_{ij}a_i\phi_i(u)r(u))^2 f(u) \times$$

$$\times du \leq 8M \int (\sum_{i,j} a_{ij} a_i \phi_j(u))^2 r(u)\, du = 8M \int [\sum_j (\sum_i a_{ij} \times$$

$$\times a_i) \phi_j(x)]^2 r(x)\, dx = 8M \sum_j (\sum_i a_{ij} a_i)^2 \leq 8M \sum_{j=1}^m a_j^2 \sum_{i,j} a_{ij}^2 \leq$$

$$\leq C_{19} \sum_{i,j} a_{ij}^2 = C_{19} \sigma_n^2. \tag{34}$$

Finally,

$$\Sigma_3 = 4(\sum_{i,j} a_{ij} a_i a_j)^2 \leq 4(\sum_{i=1}^m a_i^2)^2 (\sum_{i,j} a_{ij}^2) \leq C_{20} \sigma_n^2. \tag{35}$$

Combining the estimates for the terms Σ_1, Σ_2 and Σ_3 as given in (33), (34) and (35) we obtain from (32) the bound (30).

Next we shall assume that

$$\max_{1 \leq k \leq m} \sup_{x \in R_p} | \phi_k(x) r(x) | \leq C_{21} m^\beta, \quad \beta \geq 0.$$

In this case conditions 1° and 2° of Theorem 1 are satisfied for $\alpha = 1 + 2\beta$, $\beta \geq 0$.

Indeed,

1°
$$\sup_v \int \delta_m^2(u, v) r(u)\, du = \sup_v \int \sum_{j=1}^m \sum_{k=1}^m \phi_j(u) \phi_k(u) \times$$

$$\times \phi_j(v) \phi_k(v) r^2(v) r(u)\, du = \sup_v \sum_{j=1}^m \phi_j^2(v) r^2(v) \leq C_{22} m^\alpha,$$

$$\alpha = 1 + 2\beta.$$

2° It is easy to see that

$$\delta_{0m}(u, v) = \sum_{j=1}^m \phi_j(u) \phi_j(v) r(u) r(v).$$

Therefore

$$\sup_v \int \delta_{0m}^2(u, v) f(u)\, du \leq M \sup_v \int (\sum_{j=1}^m \phi_j(u) \phi_j(v) r(v))^2 \times$$

$$\times r(u)\, du = M \sup_v \sum_{k=1}^M \phi_k^2(v) r^2(v) \leq C_{23} m^\alpha, \quad \alpha = 1 + 2\beta,$$

and

$$\sup_v \int \delta_{0m}^4(u, v) f(u)\, du \leq C_{24} m^{2(1+2\beta)} \sup_v \int \delta_{0m}^2(u, v) f(u)\, du =$$

$$= C_{25} m^{3(1+2\beta)}.$$

Hence

$$\sup_v \int \delta_{0m}^s(u, v) f(u)\, du \leq C_{26} m^{(s-1)\alpha}, \quad \alpha = 1 + 2\beta, \quad \beta \geq 0,$$

$$s = 2; 4. \tag{37}$$

Furthermore since $\sum_{j=1}^{\infty} a_j^2 = \|f\|^2 < \infty$ from (28) and (29) we obtain that

$$\Delta_n = A_n + O(1)$$

$$\sigma_n^2 = B_n^2 + O(1), \tag{38}$$

where

$$A_n = \sum_{j=1}^{m} \int \phi_j^2(u) r^2(u) f(u)\, du,$$

$$B_n^2 = 2 \sum_{i,j}^{m} [\int \phi_j(u)\phi_i(u) r^2(u) f(u)\, du]^2.$$

Thus combining the relations derived in (30), (36), (37) and (38) we arrive at the following theorem.

THEOREM 5. Let $\max_{1\le k\le m} \sup_{x\in R_p} |\phi_j(x)r(x)| \le C_{27}m^{\beta}$, $\beta \ge 0$ and $\max_{x\in R_p} f(x)r(x) < \infty$. If $\sigma_n \to \infty$ and $n^{-1}\sigma_n^{-4}m^{3\alpha} \to 0$ $(\alpha = 1 + 2\beta)$ as $n \to \infty$, then

$$\sqrt{m}\ (\frac{n}{m} \|f_n - Ef_n\|^2 - A_{0n})B_{0n}^{-1} \xrightarrow{d} N(0,1),$$

where

$$A_{0n} = m^{-1}A_n, \quad B_{0n}^2 = m^{-1}B_n^2.$$

The conditions of Theorem 5 are simpler than those of Theorem 2.1 in Chapter 5 as well as of the theorem from [102] and [106].

We now apply this theorem to some specific orthonormal systems which were the subject of investigation by numerous authors.

1. Let a random vector X with the values in $E = [-\pi, \pi]\times[-\pi, \pi]$ be observed. It is known that this vector possesses density $f(x)$, $x \in E$. Let $\{\phi_j\}$ be a double trigonometric system of functions on E. In this case we take $(2m + 1)^2$ in place of m. It is easy to verify that

$$A_{0n} = \frac{1}{4\pi^2},$$

$$B_{0n}^2 = \frac{1}{2\pi^2} \int_E \int_E [\Phi_m(u)\Phi_n(v) f(x + u, y + v)\, dx\, dy] f(u, v) \times$$

$$\times\, du\, dv \to \sigma^2 = \frac{1}{2\pi^2} \|f\|^2,$$

where $\Phi_m(x)$ is the Fejér kernel. Thus if $m^2/n \to 0$ as $n \to \infty$ then

$$(2m + 1)\left(\frac{n}{(2m + 1)^2}\,||f_n - Ef_n||^2 - \frac{1}{4\pi^2}\right) \xrightarrow{d} N(0, \sigma).$$

2. Let $\{\phi_j\}$ be an orthonormal Legendre basis. Let $d(x) > 0$ be a bounded function on $[-1, 1]$. Assume that $f(x)$ and $d(x)$ possess bounded derivatives on $[-1, 1]$ and $d(x) = O((1 - x^2)^\gamma)$, $\gamma \geq 3/4$ for $x \uparrow 1$ and $x \downarrow -1$. Let $f_n(x)$ be the projection estimator for $f(x)\,d(x)$. In this case as it was shown in Chapter 5, Section 2,

$$A_{0n} = \int_{-1}^{1} f(x)\, d^2(x) \frac{1}{\sqrt{1 - x^2}}\, dx + O\left(\frac{\ln m}{m}\right),$$

$$B_{0n}^2 \to \sigma^2 = \frac{2}{\pi} \int_{-1}^{1} f^2(x)\, d^4(x) \frac{dx}{\sqrt{1 - x^2}} \quad \text{as} \quad n \to \infty.$$

Here $\beta = 1/2$, i.e. $\alpha = 2$. Thus if $m^4/n \to 0$ as $n \to \infty$ we have

$$m^{1/2}\left(\frac{n}{m}\,||f_n - Ef_n||^2 - \int_{-1}^{1} f(x)\, d^2(x) \frac{dx}{\sqrt{1 - x^2}}\right) \xrightarrow{d} N(0, \sigma)$$

3. Let $\{\phi_j\} = \{ P_m^{(k)}(\cos\Theta)\cos k\phi,\ P_m^{(k)}(\cos\Theta)\sin k\phi,\ k = \overline{0, m},$ $m = 1, 2, \ldots\}$ be a system of spherical functions. Let (Θ_k, ϕ_k), $k = \overline{1, n}$ be a sample from a distribution with density $f(\Theta, \phi)$ on the surface Q of a sphere of unit radius. It is required to test the hypothesis that $f(\Theta, \phi)$ is a uniform density on Q. In this case $\alpha = 3$, $A_{0n} = 1/4\pi$ and $B_{0n}^2 = 1/8\pi^2$. Thus, if $m^5/n \to 0$ as $n \to \infty$,

$$\frac{n}{m + 2} \int_Q [f_n - \frac{1}{4\pi}]^2 ds - \frac{m}{4\pi} \xrightarrow{d} N(0, \frac{1}{2\pi\sqrt{2}}).$$

The hypothesis is rejected if

$$\int_Q [f_n - \frac{1}{4\pi}]^2 ds \geq \lambda_n(\alpha),$$

where

$$\lambda_n(\alpha) = \frac{m(m + 2)}{4\pi n} + \frac{(m + 2)\sigma_0}{n}\lambda_\alpha,$$

$$\sigma_0^2 = \frac{1}{8\pi^2}, \quad 1 - \phi(\lambda_\alpha) = \alpha,$$

where $\phi(\lambda)$ is the standard normal distribution.

4. Histogram

Let $\delta_m(x, y) = m \sum_{j=1}^{m} I_j(x)I_j(y)$, where $I_j(x)$ is the indicator of the interval $[(j - 1)/m, j/m]$, $j = \overline{1, m}$. Then from (1) we obtain the histogram

$$f_n(x) = \frac{\nu_i}{nh}, \quad x \in \Delta_i = [(i - 1)/m, i/m], \quad h = \frac{1}{m},$$

where ν_i denotes the number of observations falling into the interval Δ_i, $i = \overline{1, m}$.

It is easy to observe that $\delta_m(x, y)$ is majorized by a Parzen kernel, i.e.

$$\delta_m(x, y) \leq mK(m(x - y)),$$

where $K(x)$ is an indicator of the interval $[- 1, 1]$.

Conditions 1° and 2° are fulfilled for $\alpha = 1$ since

1° $\sup_y \int \delta_m^2(x, y)\, dx \leq \sup_y \int m^2K^2(m(x - y))\, dx \leq C_{28}m,$

2° $\sup_y \int \delta_{0m}^s(x, y)f(x)\, dx \leq \sup_y \int m^sK_0^s(m(x - y))f(x) \leq C_{29}m^{s-1}$, s= 2; 4

where

$$K_0(z) = \int K(x)K(z - x)\, dx.$$

Assume that $f^{(1)}(x)$ is bounded then $m^{-1}\sigma_n^2 \to 2 \int f^2(x)\, dx$ as $n \to \infty$. Next since $\delta_m(x, y)$ is majorized by the kernel $mK(m(x - y))$, in a manner analogous to the derivation of relation (27) we prove that

$$\lambda_n = O(\sigma_n^2).$$

Consequently, condition 3° is also satisfied. Condition 4° becomes $m/n \to 0$ as $n \to \infty$. It is easy to calculate that $m^{-1}\Delta_n = 1$. Thus we state the following theorem.

THEOREM 6. If $f^{(1)}(x)$ is bounded and $m/n \to 0$ as $n \to \infty$ then

$$m^{1/2}(U_n - 1)\left(2 \int f^2(x)\, dx\right)^{-1/2} \xrightarrow{d} N(0, 1),$$

where

$$U_n = \frac{n}{m} \int (f_n(x) - Ef_n(x))^2\, dx = nm \sum_{i=1}^{m} \left(\frac{\nu_i}{n} - p_i\right)^2,$$

$$p_i = \int_{\Delta_i} f(x)\,dx, \quad i = \overline{1, m}.$$

5. Deviation of Kernel Estimators in the Sense of the Hellinger Distance

Let X_j, $j = 1, 2, \ldots$ be a sequence of independent, identically distributed random variables with values in R_1. Let their common distribution function $F(x)$ be absolutely continuous and let $f(x)$ be the corresponding probability density. As an estimator of $f(x)$ we shall consider

$$f_n(x) = \frac{m}{n} \sum_{j=1}^{n} K(m(x - X_j)),$$

where $K(x)$ is a probability density satisfying the conditions stated below and $m = m(n) \to \infty$ as $n \to \infty$.

In this section we study the limiting distribution of the deviation in the sense of the Hellinger distance ([107], [108]) of the estimators $f_n(x)$ from the density $f(x)$:

$$\rho(f_n, f) = \int (\sqrt{f_n(x)} - \sqrt{f(x)})^2\,dx.$$

Let the function $K(x)$ satisfy the following conditions: $K(x) = K(-x)$, $x^2K(x) \in L_1(-\infty, \infty)$ and $\overset{\infty}{\underset{-\infty}{V}}(K) < \infty$.

As far as $f(x)$ is concerned we shall assume that the finite interval $E_0 = [a, b]$ is its support and that it possesses derivatives up to the second order inclusively and that $f''(x)$ is bounded.

Denote

$$\Delta = (b - a) \int K^2(x)\,dx, \quad \sigma^2 = 2(b - a) \int K_0^2(u)\,du, \quad K_0 = K*K$$

($*$ is the convolution operation),

$$h_n = \inf_{x \in E_0} \sqrt{\frac{f_n(x)}{f(x)}}, \quad H_n = \sup_{x \in E_0} \sqrt{\frac{f_n(x)}{f(x)}}.$$

LEMMA 4. Let $n^{-1}m^3 \ln n \to 0$ as $n \to \infty$. Then $(h_n - 1)\sqrt{m} \overset{P}{\to} 0$ and $(H_n - 1)\sqrt{m} \overset{P}{\to} 0$ as $n \to \infty$.

<u>Proof</u>. For any $\alpha \geq 0$, $\beta > 0$ the identity

$$\sqrt{\alpha} - \sqrt{\beta} = 2^{-1}\beta^{-1/2}[(\alpha - \beta) - (\sqrt{\alpha} + \sqrt{\beta})^{-2}(\alpha - \beta)^2]$$

is valid. Therefore

$$|h_n - 1| \leq \sup_{x \in E_0} \left| \sqrt{\frac{f_n(x)}{f(x)}} - 1 \right| \leq C_{30}[\sup_{x \in E_0} |f_n(x) - f(x)| + $$
$$+ (\sup_{x \in E_0} |f_n(x) - f(x)|)^2].$$

We have

$$\sup_{x \in E_0} |f_n(x) - f(x)| \leq \sup_{x \in E_0} |f_n(x) - Ef_n(x)| + \sup_{x \in E_0} |Ef_n(x) -$$
$$- f(x)|,$$

and moreover,

$$\sup_{x \in E_0} |Ef_n(x) - f(x)| = O(m^{-2}).$$

Next applying the formula of integration by parts we obtain that

$$\sup_{x \in E_0} |f_n(x) - Ef_n(x)| \leq \sup_{x \in E_0} m \int |F_n(u) - F(u)||K(m(x -$$
$$- u))| \leq V\, m \sup_{x \in E_0} |F_n(x) - F(x)|,$$

where $V = \overset{\infty}{\underset{-\infty}{V}}(K)$ and $F_n(x)$ is the empirical distribution function based on the sample.

Hence,

$$\sqrt{m}|h_n - 1| \leq C_{31}m^{3/2} \sup_{x \in E_0} |F_n(x) - F(x)| + O(m^{-3/2}). \tag{39}$$

It follows from the inequality 1.4, in Chapter 2 of N. V. Smirnov [2] that

$$\sup_{-\infty<x<\infty} |F_n(x) - F(x)| = O_p\left(\sqrt{\frac{\ln n}{n}}\right).$$

Thus we obtain from (39) that

$$\sqrt{m}|h_n - 1| = O_p\left(\sqrt{\frac{m^3 \ln n}{n}}\right) + O(m^{-3/2}).$$

Hence $\sqrt{m}(h_n - 1) \overset{p}{\to} 0$ as $n \to \infty$. The convergence $\sqrt{m}(H_n - 1) \overset{p}{\to} 0$ as $n \to \infty$ is proved analogously.

THEOREM 7. Let $nm^{-9/2} \to 0$ and $(m^3 \ln n)/n \to 0$ as $n \to \infty$. Then

$$\sqrt{m}\,(L_n - \Delta_0)\sigma_0^{-1} \xrightarrow{d} N(0,\,1),$$

where

$$L_n = \frac{n}{m}\,\rho(f_n,\,f), \quad \Delta_0 = \frac{\Delta}{4}, \quad \sigma_0^2 = \frac{\sigma^2}{16}\,.$$

Proof. Observe that

$$U_n = \frac{n}{m}\int (f_n(x) - f(x))^2\,\frac{dx}{f(x)} = \frac{n}{m}\int (\sqrt{f_n(x)} - \sqrt{f(x)})^2 \times$$

$$\times \left(1 + \sqrt{\frac{f_n(x)}{f(x)}}\right)^2 dx.$$

Whence

$$(1 + H_n)^{-2}U_n \le L_n \le (1 + h_n)^{-2}U_n \tag{40}$$

Furthermore we obtain from (40) that

$$\sqrt{m}\,(L_n - \Delta_0)\sigma^{-1} \le m^{1/2}(1 + h_n)^{-2}(U_n - \Delta)\sigma^{-1} + m^{1/2}\Delta\sigma^{-1} \times$$

$$\times[(1 + h_n)^{-2} - \tfrac{1}{4}]. \tag{41}$$

Utilizing Theorem 3 and the fact that $h_n \xrightarrow{p} 1$ which follows from Lemma 4 we have

$$m^{1/2}(1 + h_n)^{-2}\sigma^{-1}(U_n - \Delta) \xrightarrow{d} N(0,\,\tfrac{1}{4}) \tag{42}$$

Next it is easy to verify that

$$\sqrt{m}\,\Delta\sigma^{-1}|(1 + h_n)^{-2} - \tfrac{1}{4}| \le C_{32}(|h_n - 1| + |h_n - 1|^2)\sqrt{m}.$$

Hence, in view of Lemma 4 the second summand in (41) converges to zero in probability.

Thus from here and relations (41) and (42) we arrive at the inequality

$$\underline{\lim}_{n\to\infty}\; P\{\sqrt{m}(L_n - \Delta_0)\sigma^{-1} < \lambda\} \ge \Phi(4\lambda).$$

The inequality

$$\overline{\lim}_{n\to\infty}\; P\{\sqrt{m}\,(L - \Delta_0)\sigma^{-1} < \lambda\} \le \Phi(4\lambda),$$

is proved in an analogous manner. Here $\Phi(x)$ is the distribution function of the standard normal law.

Whence

$$\sqrt{m}\,(L_n - \Delta_0)\sigma_0^{-1} \xrightarrow{d} N(0,1)$$

and the theorem is proved.

Remark. An analogous result can be obtained in the same manner for deviations of estimators $f_n(x)$ from the density $f(x)$ in the sense of the Kullbuck-Leibler distance.

We now pose the following problem: Is there an almost sure convergence of the sequence $\{f_n(x)\}$, $n = 1, 2, \ldots$ to $f(x)$ in the sense of the Hellinger distance without any assumptions on the density $f(x)$?

THEOREM 8. Let $K(x)$ be a bounded probability density and the series $\sum_{n=1}^{\infty} ne^{-\gamma\lambda_n}$, $\lambda_n = nm^{-2}$ be convergent for all $\gamma > 0$. Then for any density $f(x)$ as $n \to \infty$

$$\int (\sqrt{f_n(x)} - \sqrt{f(x)})^2\, dx \to 0$$

almost surely.

This theorem is a simple corollary of Theorem 3.1 in Chapter 2.

REFERENCES

Translator's Remark:

The references presented below are consecutively numbered and referred to by those numbers in the text. Books and journal articles by Soviet authors that have been translated into English are cited as they appear in the English translation. Also the original English language editions of books (which have been translated into Russian) are cited.

1. Glivenko, V. I. (1934) Course in Probability Theory, Moscow (in Russian).
2. Smirnov, N. V. (1951) On the approximation of probability densities of random variables, Scholarly Notes of Moscow State Polytechnical Institute 16, 3, pp. 69-96 (in Russian). Also: Theory of Probability and Mathematical Statistics, Collected Works, Nauka, Moscow, 1970, pp. 205-233 (in Russian).
3. Tumanyan, S. Kh. (1955) On maximal deviation of empirical probability density. Scholarly Works of the Erevan State University, 48, pp. 3-48 (in Russian).
4. Rosenblatt, M. (1956) Remarks on some non-parametric estimates of a density function. Annals Math. Statist., 27, 3, pp. 832-837.
5. Parzen, E. (1962) On estimation of a probability density function and mode. Annals Math. Statist., 33, 3, pp. 1065-1076.
6. Chenčov, N. N. (1962) Estimation of unknown probability density based on observations, Dokl. Akad Nauk SSSR, 147, I, pp. 45-48 (in Russian).
7. Loftsgarden, D., Quessenberry C. P. (1965) A nonparametric estimate of multivariate density function, Annals Math. Statist., 36, pp. 1049-1051.
8. Maniya, G. M. (1974) Statistical Estimation of Probability Distributions, Tbilisi (in Russian).
9. Schuster, E. (1970) Note of the uniform convergence of density estimates, Ann. Math. Statist., 41, pp. 1347-1348.
10. Collomb, G. (1978) Estimation non parametrique de la regression, Université Paul Sabatier Toulouse, pp. 70-78.
11. Chentsov, N. N. (1980) On statistical estimates of an unknown probability distribution law, Theor. Probl and Applic. 25, 3, pp. 363-637). (paper presented at the seminar on Mathematical Statistics commemorating L. N. Bol'shev, 8 Oct. 1979, Math. Institute of the Academy of Sciences of the USSR).
12. Devroye, P. (1978) The uniform convergence of the Nadaraya-Watson regression function estimate. The Canadian Journal of Statistics, 6, 2, pp. 179-191.
13. Farrel, R. (1972) On the best obtainable asymptotic rates of convergence in estimation of a density function at a point. Ann. Math. Statist., 43, 1, pp. 170-180.

·14. Woodroofe, M. (1970) On choosing a delta-sequence. Ann. Math. Statist., 41, 5, pp. 1665-1671.
15. Van Ryzin, J. (1969) On strong consistency of density estimates. Ann. Math. Statist., 40, pp. 1765-1772.
16. Vaduva, I. (1968) Contributions to the theory of statistical estimation of probability density functions and applications. Stud.Cerc. Mat., 20, pp. 1207-1276.
17. Schuster, E. (1969) Estimation of a probability density function and its derivatives. Ann. Math. Statist., 40, pp. 1187-1195.
18. Blumenthal, S. (1963) The asymptotic normality of two test statistics associated with the two-sample problem. Ann. Math. Statist., 34, 4, pp. 1513-1523.
19. Orlov, A. I. (1972) On testing the symmetry of distributions, Theor. Prob. and Applic., 17, 2, pp. 357-361.
20. Butler, C. (1969) A test for symmetry using the sample distribution function, Ann. Math. Stat., 40, pp. 2209-2210.
21. Nadaraya, E. A. (1964) On estimating regression, Theor. Probab. and Applic., 9, pp. 141-142.
22. Stone, G. (1977) Consistent nonparametric regression, Ann. Math. Statist., 5, pp. 595-645.
23. Watson, G. (1964) Smooth regression analysis, Sankhyâ, Ser. A, 26, pp. 359-372.
24. Konakov, V. D. (1974) Some Problems of Nonparametric Estimation in Multidimensional Statistical Analysis, Doctoral Dissertation, Moscow (in Russian).
25. Bernstein, S. N. (1941) On fiducial 'Fisher probabilities', Izvest. Akad Nauk SSSR, Ser. Mat., 5, pp. 85-93 (in Russian).
26. Rooney, P. (1957) On the inversion of the Gauss transformation, Canadian Journal of Math., 9, pp. 125-134.
27. Watson, G. and Leadbetter, M. (1963) On the estimation of the probability density, Ann. Math. Statis., 34, pp. 480-491.
28. Bartlett, M. (1963) Statistical estimation of density functions, Sankhyâ, Ser. A., 25, pp. 245-254.
29. Von Bahr, B. (1965) On the convergence of moments in the central limit theorem. Ann. Math. Statist., 36, 3, pp. 808-818.
30. Petrov, V. V. (1962) On certain polynomials arising in probability theory. Vestnik Leningrad University, 19, 4, pp. 150-133 (in Russian).
31. Nikol'skii, S. M. (1969) Approximation of Functions of Several Variables and Embedding Theorems, Nauka, Moscow (in Russian).
32. Nadaraya, E. A. (1969) Estimation of Density of Bivariate Distributions, Nauka, Moscow (in Russian).
33. Cacoullos, T. (1966) Estimation of a multivariate density, Ann. Inst. Stat. Math., 18, 2, pp. 179-189.
34. Mirzakhmedov, M. A. (1969) On estimation of multidimensional probability density, Scholarly Works of the Tashkent State University, Tashkent, pp. 146-150 (in Russian).
35. Smirnov, N. V. (1944) Approximation of the probability distributions of random variables from empirical data, Uspekhi Matem. Nauk, 10, pp. 179-206 (in Russian).
36. Kiefer, J. and Wolfowitz, J. (1958) On the deviations of the empiric distribution function of vector chance variables, Trans. Amer. Math. Soc., 87, 1, pp. 173-186.

37. Prokhorov, Yu. V. (1968) An extension of S. N. Bernstein's inequalities to multidimensional distributions, Theory Probab. and Applic., 13, 2, pp. 260-267.
38. Hoeffding, W. (1948) A class of statistics with asymptotically normal distribution, Ann. Math. Statistics, 19, pp. 293-325.
39. Sheffé, H. (1947) A useful convergence theorem for probability distributions, Ann. Math. Stat., 18, 3, pp. 434-438.
40. Aivazyan, S. A., Bezhaeva, Z. I. and Staroverov, O. V. (1974) Classification of Multivariate Observations, Moscow (in Russian)
41. Van Ryzin, J. (1966) Bayes risk consistency of classification procedures using density estimation, Sankhyà, Ser. A., 28, pp. 261-270.
42. Cramér, H. (1944) On a new limit theorem of probability theory; Russian version in: Uspekhi Mat. Nauk, 10, pp. 166-178.
43. Sazanov, V. V. (1968) On the speed of convergence in the multidimensional central limit theorem, Theor. Probab. and Applic., 13, pp. 188-191.
44. Hobson, E. (1927) The Theory of Functions of a Real Variable and the Theory of Fourier's Series, vol. I, Section 351, Cambridge Univ. Press.
45. Smirnov, N. V. (1939) On deviations of the empirical distribution function, Mat. Sbornik, 6, pp. 3-24 (in Russian).
46. Mises, R. (1921) Das problem der iteration. Zeitschrift für Angew. Math. und Mech., 4, 1, pp. 298-307.
47. Smirnov, N. V. (1937) On a number of sign changes in a sequence of deviations. Izv. AN SSSR, 3, pp. 361-371.
48. Cramér, H. (1946) Mathematical Methods of Statistics, Princeton University Press, Princeton, N.J.
49. Bickel, P. and Rosenblatt, M. (1973) On some global measures of the deviations of density function estimates, Ann. Statis., 1, 6, pp. 1071-1095.
50. Konakov, V. D. (1978) Complete asymptotic expansions for the maximum deviation of the empirical density function, Theor. Probab. and Applic., 23, No. 3, pp. 475-489.
51. Komlos, J., Major, P. and Tusnady, G. (1975) An approximation of partial sums of independent RV-S, and the sample DF.I, Z. Wahrscheinlichkeitstheor.Verw.Geb., 32, 1 - 2, pp. 111-131.
52. Brillinger, D. (1969) An asymptotic representation of the sample distribution function, Bull. Amer. Math. Soc., 75, pp. 545-547.
53. Gikhman, I. I. and Skorohod, A. V. (1979) Theory of Stochastic Processes I, Springer-Verlag, N.Y. (translated by S. Kotz).
54. Tumanyan, S. Kh. (1956) Asymptotic distribution of the χ^2-criterion when the number of observations and the number of groups increases simultaneously, Theor. Probab. and Applic., 1, pp. 117-132.
55. Csörgo, M. (1979) Strong approximations of the Hoeffding, Blum, Kiefer, Rosenblatt multivariate empirical process, J. Multivariate Analysis, 9, pp. 84-100.
56. Rosenblatt, M. (1975) A quadratic measure of deviation of two-dimensional density estimates and a test of independence, Ann. Statist., Statist., 3, pp. 1-14.
57. Bol'shev, L. N. and Smirnov, N. V. (1965) Tables of Mathematical Statistics, Nauka, Moscow, pp. 81-83.
58. Leonov, V. P. and Shiryaev, A. N. (1959) On a method of calculating semi-invariants, Theor. Probab. and Apllic., 4, 3, pp. 319-329.
59. Hoeffding, W. and Robbins, H. (1948) The central limit theorem for dependent random variables, Duke Math. J., 15, pp. 773-780.

60. Konakov, V. D. (1977) On a global measure of deviation for an estimate of the regression line, Theor. Probab. and Applic., 22, 4, pp. 858-868.
61. Revesz, P. (1976) On strong approximation of the multidimensional empirical process, Ann. Probab., 4, pp. 729-743.
62. Rao, C. R. (1965) Linear Statistical Inference and its Applications. J. Wiley, N.Y.
63. Watson, G.S. (1969) Density estimation by orthogonal series, Ann. Math. Statist., 40, 4, 1496-1498.
64. Schwartz, S. G. (1967) Estimation of probability density by an orthogonal series, Ann. Math. Stat., 38, 4, pp. 1261-1265.
65. Khashimov, Sh. A. (1979) Estimation of a probability density by Laguerre polynomials. In: Sluchainye Protsessy i Statistich. Vyvody, issue III (1981), pp. 186-192, Nauka, Moscow (in Russian).
66. Ibragimov, I. A. and Khasminskii, R. Z. (1979) Statistical Estimation - Asymptotic Theory, Springer-Verlag, N.Y. (translated by S. Kotz).
67. Zygmund, A. (1952) Trigonometric Series, Chelsea N.Y. 2nd ed.
68. Kronmal, R. and Tarter, M. (1968) The estimation of probability densities and cumulatives by Fourier series methods, J. Amer. Statist. Assoc., 63, pp. 925-952.
69. Grenander, U. and Szegő, G. (1958) Toeplitz Forms and their Applications, Berkeley, Univ. of California Press.
70. Suyetin, P. K. (1976) Classical Orthogonal Polynomials, Nauka, Moscow (in Russian).
71. Chibisov, D. M. and Gvantseladze, L. G. (1975) On goodness of fit tests based on grouped data. In: Third Soviet-Japanese Symposium on Probability Theory. Fan, Tashkent, 1975, pp. 183-185 (in Russian).
72. Nadaraya, E. A. (1965) On non-parametric estimates of density functions and regression curves, Theor. Prob. and Applic., 10, I, pp. 186-190.
73. Nadaraya, E. A. (1965) Nonparametric estimation of regression curves, Transact. Comp. Center AN Georg SSR, 5, 1, pp. 56-68 (in Russian).
74. Nadaraya, E. A. (1969) On nonparametric estimation of derivatives of probability density and regression function. Proc. AN Georg SSR, 55, 1, pp. 29-32 (in Russian).
75. Nadaraya, E. A. (1970) Remarks on nonparametric estimates for density functions and regression curves, Theor. Prob. and Applic., 15, 1, pp. 134-137.
76. Nadaraya, E. A. (1970) On the construction of confidence regions for probability densities, Proc. AN Georg SSR, 59, 1, pp. 33-36 (in Russian).
77. Nadaraya, E. A. (1972) On the mean square error of some nonparametric estimators of probability density, Proc. AN Georg. SSR, 67, 2, pp. 289-292 (in Russian).
78. Nadaraya, E. A. (1972) On the integral mean square error of some nonparametric estimators of probability density, Proc. AN Georg SSR, 68, 1, pp. 33-36 (in Russian).
79. Nadaraya, E. A. (1972) On the construction of confidence regions for probability density, Abstracts of Lectures at the Seminar of the Vekua Institute of Applied Mathematics, Tbilisi State University, 6, pp. 27-32 (in Russian).
80. Nadaraya, E. A. (1973) On convergence in the L_1 norm of probability density estimates, Theor. Prob. and Applic., 18, pp. 808-811.

81. Nadaraya, E. A. (1973) On the integral mean square error of some nonparametric estimators of probability density, Abstracts of papers at the International Vil'nyus Conference on Probability Theory and Mathematical Statistics, Vil'nyus (in Russian).
82. Nadaraya, E. A. (1973) Some limit theorems related to nonparametric estimates of regression functions, Proc. AN Georg. SSR, 71, 1, pp. 56-60 (in Russian).
83. Nadaraya, E. A. (1974) On the mean square error of some nonparametric estimators of probability density, Theor. Prob. and Math. Statistics, 10, pp. 116-129 (in Russian).
84. Nadaraya, E. A. (1974) Limiting distribution of quadratic deviation of nonparametric regression estiamtors, Proc. AN Georg SSR, 74, 1, pp. 33-36 (in Russian).
85. Nadaraya, E. A. (1974) On the integral mean square error of some nonparametric estimates for the density function, Theor. Prob. and Applic., 19, 1, pp. 133-141.
86. Nadaraya, E. A. (1975) On the construction of confidence regions for probability densities, Transact. of Vekua Institute of Applied Mathematics, Tbilisi State University, pp. 183-204.
87. Nadaraya, E. A. (1975) Limiting distribution of quadratic deviation of two nonparametric density estimators, Proc. AN Georg SSR, 78, 1, pp. 25-28.
88. Nadaraya, E. A. (1975) On testing symmetry of a distribution. Proc. AN Georg SSR, 78, 2, pp. 277-280.
89. Nadaraya, E. A. (1975) On homogeneity tests related to nonparametric density estimators, Third Soviet-Japanese Symposium, Abstracts of Papers Tashkent, pp. 121-123 (in Russian).
90. Nadaraya, E. A. (1976) On the nonparametric estimator of Bayesian risk in the classification problem, Proc. AN. Georg SSR, 82, 2, pp. 277-280 (in Rusisan).
91. Nadaraya, E.A. (1976) On the quadratic measure of deviation of the projection estimate of a distribution density, Theor. Prob. and Applic., 21, 4, pp. 843-850.
92. Nadaraya, E. A. (1977) Limiting distribution of the quadratic deviation of nonparametric regression estimators, Second Vil'nyus Conference on Probability Theory and Mathematical Statistics, Abstracts of Papers, pp. 70-73 (in Russian).
93. Nadaraya, E. A. (1979) On maximal deviation of nonparametric estimators of probability density in some degenerate cases, Proc. AN Georg SSR, 93, 3, pp. 561-564 (in Russian).
94. Nadaraya, E. A. (1979) On the maximal deviation of nonparametric kernel type density estimates in some degenerate cases, 12th European Meeting of Statisticians, Varna, page 171.
95. Nadaraya, E. A. (1980) Remarks on the convergence of kernel-type density estiamtors in variation, Proc. AN Georg SSR, 97, 3, pp. 577-580 (in Russian).
96. Nadaraya, E. A. (1980) On Rosenblatt's test for independence, Proc. AN Georg SSR, 98, 1, pp. 33-36.
97. Nadarya, E. A. (1980) On some tests based on kernel type density estimators, Proc. AN Georg SSR, 99, 1, pp. 37-40.
98. Nadaraya, E. A. (1980) Some problems in nonparametric estimation of probability densities and a regression curve, Theor. Prob. and Applic., 25, 3, pp. 637-638 (Paper presented at the seminar on Math. Statistics commemorating L. N. Bol'shev, 8 Oct. 1979, Steklov Math. Institute of the USSR Academy of Sciences).

99. Nadaraya, E. A. (1980) On maximal deviation of the kernel type non-parametric density estimates in some degenerate cases, Math. Operations forsch. Statist., Ser. Statistics, 11, 4, pp. 1-16.
100. Nadaraya, E.A. (1984) Application of a martingale central limit theorem for a study of the limiting distribution of quadratic deviation of the kernel-type density estimators. Proc. AN George SSR, 113, 2, pp. 253-256 (in Russian).
101. Hall, P. (1984) Central limit theorem for integrated square error of multivariate non-parametric density estimators. Journal of Multivariate Analysis, 14, pp. 1-16.
102. Ghorai, J. (1980) Asymptotic normality of a quadratic measure of orthogonal series type density estimate. Annals of the Institute of Stat. Math., 32, part A, pp. 341-350.
103. Liptser, R. Sh. and Shiryaev, A. N. (1980) A functional central limit theorem for semimartingales, Theor. Prob. and Applic., 25, 4, pp. 669-688.
104. Nadaraya, E. A. and Absava, R. M. (1985) On the limiting distribution of the quadratic deviation of probability density estimators. IV Intern. Vilnyus Conf. on Theor. Prob. and Math. Stat., Abstracts (in Russian).
105. Absava, R. M. (1985) On the limiting distribution of quadratic deviation for a wide class of probability density estimators. Proc. AN Georg SSR, 120, 1, pp. 45-48 (in Russian).
106. Buadze, T. G. (1984) On the limiting behavior of projection probability density estimators. Proc. AN Georg SSR, 104, 3, pp. 555-558 (in Russian).
107. Beran, R. (1977) Minimum Hellinger distance estimates for parametric models, Ann. Stat., 5, 3, pp. 445-463.
108. Absava, R. M. (1985) Limiting distribution in the sense of Hellinger's distance of kernel probability density estimators, Proc. AN Georg SSR, 120, 3, pp. 477-480 (in Russian).

AUTHOR INDEX

Aivazyan, S. A., 206
Absava, R. M., 177, 209

Bartlett, M., 19, 205
Brillinger, D., 89, 206
Bol'shev, L. N., 206
Bernstein, S. N., 4, 152
Blumenthal, S. R., 205
Butler, G. A., 205
Bezhaeva, Z. I., 206
Bickel, P., 206
Buadze, T. G., 209
Bezan, R., 209

Chentsov, N. N., 1-3, 7, 17, 57, 161, 162, 194, 204
Chibisov, D. M., 207
Csõrgo, M., 206
Cramér, H., 10, 64, 102, 206
Cocoullos, T., 205

Devroye, P., 204

Fazzel, R., 204

Glivenko, V. I., 1, 204
Gikhman, I. I., 206
Grenander, U., 207
Gvantseladze, L. G., 207
Ghozai, J., 177, 209

Hoeffding, W. A., 51, 52, 147, 206
Hobson, E., 206
Hall, P., 177, 209

Ibzagimov, I. A., 162, 207

Kolmogorov, A. N., 10, 102
Kendall, M. G., 85
Khas'minskiĭ, R. Z., 162, 207
Kronmal, R., 169, 207
Komlos, I., 206
Konakov, V. D., 205-207
Kiefer, J. 205
Khashimov, Sh. A., 207

Loftsgarden, D., 204
Leonov, V. P., 206
Liptser, R. Sh., 177, 209

Maniya, G. M., 2, 204
Melamed, I. A., 17
Mises, R., 10, 72, 99, 206
Mirzakhmedov, M. A., 205
Major, P., 206

Nikol'skiǐ, S. M., 205
Nadaraya, E. A., 177, 205, 207-209

Orlov, A. I., 205

Prohorov, Yu. V., 49, 50, 206
Parzen, E., 1-2, 42, 186, 192, 204
Petrov, V. V., 25, 205

Rosenblatt, M., 1, 2, 18, 20, 186, 192, 204, 206
Revesz, P., 207
Rao, C. R., 207
Rooney, P., 205
Robbins, H., 206

Smirnov, N. N., 1, 6, 7, 11, 16, 42, 43, 62, 72, 201, 204-206
Skorohod, A. V., 206
Sazonov, V. V., 67, 206
Stuart, A., 85
Shiryaev, A. N., 177, 206, 209
Schwartz, S. G., 207
Szego, G., 207
Suyetin, P. K., 207
Stone, G., 205
Sheffe, H. A., 55, 206
Schuster, E., 6, 205
Starovarov, O. V., 206

Tumanyan, S. Kh., 1, 204, 206

Tarter, M., 169, 207
Tusnady, G., 206

Van Ryzin, 6, 205, 206
Vaduva, I., 6, 205
Von Bahr, 36, 205

Wolfowitz, J., 205
Woodroofe, M., 205
Watson, G., 205, 207

Zygmund, A., 207

SUBJECT INDEX

Asymptotically optimal estimators, 3
Asymptotical normality, 9
Asymptotic confidence region, 13
Asymptotic test, 95, 151
Asymptotic power, 99
Asymptotic confidence interval, 160

Bayesian classification rule, 57
Bayesian decision rule, 58
Brownian bridge, 87, 110
Bahr's formula, 146
Borel-Cantelli's Lemma, 163
Brownian motion process, 164
Berry-Essen's theorem, 37
Breiman-Brillinger approximation, 177

Consistent estimator, 26
Convergence:
- in the L_2 norm, 47
- in variation, 54
Convolution, 49
Characteristic function, 112, 144, 167
Confidence region:
- for densities, 80, 85
- for regression curve, 134, 135
Classification problem, 54
Chebyshev-Hermite polynomial, 25
Conjugated distribution function, 65
Chentsov estimators, 194

Dirichlet
- series, 44
- Kernel, 169
Delta measure, 1
Density estimators of a general type, 177
Difference-martingale, 179
Duuble trigonometric system of functions, 197

Empirical distribution, 1
Estimator Chentsov's type, 3
Empirical distribution function, 43, 87
Empirical classification region, 57
Empirical process, 87, 110, 164
Estimation:
- of density, 2, 3, 18
- of derivative of the density, 35, 152
- multivariate case, 21, 30
- of mode, 44
- of regression function, 115
- of density projection type, 161
- of components of a convolution, 152
Examples of the Kernel, 176

Feje'r integral, 169
Fourier coefficients, 173
Fubini's theorem, 19

Goodness of fit test, 16
Generalized Minkowski inequality, 29

Histogram, 76, 132, 199
Hõlder's inequality, 145
Hoeffding theorem, 147
Hermit polynomial, 154
Hellinger distance, 200

Integrated mean square error, 18
Independence test, 109

Kernel type probability density estimator, 18, 186
Kullbuck-Leibler distance, 203

Limiting
- power function, 11
- distribution, 62
Local behavior of power, 97

Limiting distribution:
- of maximal deviation of Kernel type estimator, 62
- of the maximal deviation of estimator of regression curve, 125
- of the squared norm of a projection type density estimator, 164
- of quadratic deviation of estimator of regression curve, 138
- of quadratic deviation for a wide class of probability density estimator, 177
- of the deviation in the sense of Hellinger distance of Kernel estimator, 200
Legendre polynomials, 170
Lebesque constant, 169

Mode, 2, 44
Mean integral squared error, 4, 18
Maximal deviation, 7, 125
Mean square error, 30
Maximum likelihood method, 54
Multivariate central limit theorem, 120
Martingale method, 177

Nonparametric density estimation, 1, 18
Normal distribution, 37

Optimal estimator of the density, 35
Orthonormal Legendre basis, 198

Projection type estimation, 2, 194
Poisson process, 142
Parseval's inequality, 168

Quadratic deviation:
- of two nonparametric Kernel type density estimators, 87
- of estimating an unknown regression curve, 120, 138
- projection type density estimator, 164

Regression curve, 12, 115
Robbins theorem, 147
Rao's theorem, 160
Rosenblatt-Parzen estimators, 186

Sample made, 6, 44
Singular alternatives, 11, 99, 108, 193
Sequences of estimators, 47
Standard Wiener process, 87, 164
Strong consistency:
- of Kernel type density estimators, 42
- of regression curve estimators, 122
- of projection type density estimator, 161
Sheffe's theorem, 55
Stochastic sequence, 179
System of spherical functions, 198

Testing hypothesis:
- for symmetry, 103
- for independence, 109
- for regression curve, 138
- for density, 96
- about equality of two regression functions, 151
Toeplitz form, 170
Two-dimensional Brownian bridge, 110

Uniform distance, 48